Electric & Hybrid Vehicles

A Technology towards the historical development of the Automobile industry

Shubhangi S. Joshi

notionpress.com
INDIA • SINGAPORE • MALAYSIA

ISBN
Paperback 979-8-90392-564-3
Hardcase 979-8-90392-565-0

Gratitude

The significance of Electric and Hybrid vehicles is now recognized across the globe, influencing diverse engineering disciplines and driving the transition toward sustainable mobility. With the rising demand for clean transportation, this subject has become both relevant and essential for today's learners. Motivated by the need for clear and concise academic material, I have developed this book to present the fundamentals in a structured, engaging, and student-centered manner. The content is written in a simple and coherent style, breaking down complex concepts into easy, digestible steps. Labelled diagrams, illustrations, and real-world case studies support visual clarity and practical understanding. Each chapter is arranged in a logical progression, ensuring that concepts connect naturally and enhance the overall learning flow. To reinforce comprehension, key points and representative questions are provided at the end of every section. These not only support exam preparation but also deepen conceptual insight. The aim is to offer a complete, accessible resource that covers the syllabus thoroughly without burdening the reader with unnecessary complexity.

I hope this book proves valuable for both students and educators—helping learners master essential concepts while encouraging them to explore the broader landscape of electric and hybrid mobility.

My heartfelt gratitude goes to everyone who supported the creation of this work. A formal acknowledgment and gratitude to the segment of the Bibliography & Reference section.

A special note of appreciation is extended to Neotech Institute of Technology for shaping my academic and professional journey. I also wish to acknowledge my family, especially my spouse, and loving daughters, for their constant encouragement, and last but not least, my design team for their dedication to timely and high-quality production.

Constructive feedback and suggestions for improvement are always welcome and sincerely appreciated.

Author **Shubhangi S. Joshi**
(M.E. E&TC)

Preface

A quick dive into the history of Electric and Hybrid Vehicles, tracing their journey from early prototypes to today's high-efficiency, eco-driven mobility tech. The content also breaks down the major drivetrain topologies, giving a clear view of how different EV and HEV configurations channel and manage power. Alongside this, the control mechanism approaches are explained in a smooth, easy-to-grasp way, showing how modern systems balance performance, efficiency, and real-time energy flow. Electric vehicles run solely on battery power, while hybrid vehicles combine an internal combustion engine with an electric motor and battery. Plug-in hybrids are a type of hybrid that can be charged from an external power source, allowing for a greater electric-only range than a standard HEV. The primary goal of both is to improve fuel economy and reduce emissions compared to traditional gasoline-only cars. The transformation from a Battery Electric Vehicle to a Hybrid Electric Vehicle involves evolving a purely electric powertrain into a dual-source system that combines electric propulsion with an internal combustion engine.

In this book, I have focused on making concepts as clear and student-friendly as possible. Detailed diagrams, graphical representations, and structured comparisons are used throughout to help readers easily distinguish between the different types of electric vehicles. I aim to support learners in strengthening their understanding, building real technical expertise in EV and HEV systems, and ultimately enhancing their academic performance. I hope this book becomes a valuable resource in their journey toward mastering the field.

Electric and Hybrid Vehicles

A Technology towards the historical development of the Automobile Industry

The evolution of the automobile has undergone a significant transformation with the emergence and re-emergence of electric and hybrid electric vehicles (EVs and HEVs). Although electric vehicles appeared as early as the late nineteenth century, their prominence declined with the rise of affordable internal combustion engine vehicles, most notably after the introduction of the Ford Model T in 1908. Renewed interest in EVs during the 1970s oil crisis and the environmental movements of the 1990s, combined with recent advancements in battery technology and power electronics, has positioned electric and hybrid vehicles as viable and sustainable alternatives to conventional transportation.

The information presents a comprehensive study of electric and hybrid vehicle technologies, focusing on system architecture, energy storage, electric drive trains, power electronics, vehicle dynamics, and energy management strategies. The work begins with an examination of the historical development and adoption of EVs, highlighting technological milestones, economic factors, and regulatory influences that have shaped modern electric mobility. Emphasis is placed on the growing role of smart grids and nationwide charging infrastructure in enabling large-scale EV deployment. Fundamental principles of vehicle motion and dynamics are analyzed to understand performance characteristics, including rolling resistance, torque-speed behavior, and efficiency limitations. Various types of electric vehicles—including battery electric vehicles (BEVs), hybrid electric vehicles (HEVs), plug-in hybrids (PHEVs), fuel cell electric vehicles (FCEVs), and solar electric vehicles—are classified and compared based on operating principles, advantages, and suitability for specific applications. A detailed discussion of electric motors, including DC, brushless DC, induction, permanent magnet, and switched reluctance motors, provides insight into their control strategies and performance in traction applications. The contents further explore hybrid electric vehicle architectures and drive-train topologies, with particular focus on parallel and power-split configurations. Power flow control, regenerative braking systems, communication networks, and the impact of modern drive-trains on energy consumption are critically evaluated. Energy storage systems form a central component of the study, covering battery-based systems, fuel cells, flywheels, and hybrid energy storage configurations. Comparative analysis of lithium-ion and lead-acid batteries highlights efficiency, performance, and lifecycle considerations. Optimization-based energy management systems and charging techniques, including grid-connected and solar-powered charging stations, are discussed as key enablers of efficient vehicle operation.

Finally, the book addresses future trends in electric mobility, including vehicle-to-grid integration, smart energy systems, advanced power electronics, artificial intelligence applications, and alternative hydrogen fuel technologies. These developments underscore the critical role of electric and hybrid vehicles in achieving sustainable transportation and energy resilience in the twenty-first century.

– Author

Table of Contents

Chapter 01

Historical Development

1.1 Evolution of EV

The evolution of electric vehicles (EVs) began between 1828 and 1835, when innovators like Ányos Jedlik and Thomas Davenport created early small-scale prototypes. A major milestone came in 1881, when Frenchman Gustave Trouve built the first functional EV—it is a tricycle powered by a 0.1-horsepower DC motor, using a lead-acid battery, and weighing around 160 kg. By the late 1800s and early 1900s, practical electric vehicles gained traction in the US and Europe due to their clean, quiet operation. During this time, Morris and Salom's electro boat emerged as the first commercially available electric vehicle, operating as a hired mode of transport in New York City. Despite its higher initial investment, it proved to be more profitable than traditional horse-drawn cabs. It ran three 4-hour shifts daily with 90-minute recharges, well equipped with dual 1.5 hp motors. This system can achieve speeds of up to 32 km/h and provides a maximum range of 40 km. Between 1910 and 1930, the large-scale manufacturing of internal combustion engine vehicles, such as the Ford Model T, resulted in a significant decline in the use of electric vehicles. Notably, M.A. Darracq's 1897 coupe introduced regenerative braking, allowing recovery of kinetic energy during deceleration to recharge the batteries, significantly improving efficiency. One of the era's standout vehicles was La Jamais Contente by Camille Jenatzy, which holds the distinction of being the first automobile surpasses speed of 100 kilometres per hour. Later, in 1966, General Motors (GM) developed the Electro van, using induction motors and thyristor-based inverters—paving the way for modern EV architecture. The 1970s oil crisis revived interest in EVs due to fuel shortages, though technological limitations remained. In the 1990s, environmental concerns and regulatory mandates led to new developments like GM's EV1. The modern era began in 2008 with the launch of the Tesla Roadster, showcasing the power of lithium-ion technology. Since then, global automakers have entered the EV market, making rapid advancements in battery efficiency, driving range, and charging infrastructure—positioning EVs at the forefront of sustainable transportation.

Fig.1.1: Battery Electric Vehicle transformation

***Fact:** In 1900, about 38% of all cars in the US were electric! Electric vehicles were very popular in urban areas.*

1.2 Development Towards The 21st Century

As the 21st century approached, the development of electric vehicles (EVs) accelerated significantly because of several technological, regulatory, and economic factors. One of the most critical enablers was the advancement in battery technologies, particularly the commercialization of lithium-ion batteries, and ongoing research into solid-state batteries, which promise greater energy density, faster charging, and improved safety. Parallel to technological improvements, environmental regulations such as the

European Union's CO_2 emissions targets and the United States Corporate Average Fuel Economy (CAFÉ) standards placed pressure on automakers to reduce emissions and improve fuel efficiency, indirectly encouraging the shift toward electric mobility.

Historical Development of Automobile GTUexam-3170923-SUM2025

The historical evolution of the automobile, along with the growing interest and activity in electric vehicles from 1890 to the present day, is noteworthy for several advantages. The automobile era began in the late 1800s, with electric vehicles (EVs) enjoying a golden age in the early 1900s before being surpassed by the more affordable gasoline-powered Ford Model T in 1908. Interest in EVs resurged in the 1970s due to the oil crisis, and again in the 1990s with environmental regulations and improved battery technology. Since the late 20th century, advancements in batteries, increased consumer choice, and environmental concerns have driven modern EV development, creating a nationwide charging infrastructure and achieving widespread commercial availability, with models like the Nissan LEAF and significant drops in battery costs.

- **Early Concepts (Late 1800s):** The development of practical, commercially available electric, steam, and gasoline-powered vehicles occurred in the late 19^{rh} century.
- **The First Automobile Era (1880s-1900s):** Early innovators in France, England, and the U.S. created the first electric cars and other motor vehicles.
- **The Gasoline Dominance (Early 1900s):** This, introduced in 1908, made gasoline cars accessible to the masses through affordability and mass production. This led to the construction of a widespread network of gas stations, further solidifying gasoline's dominance.
- **A Century of Gasoline Cars (1920s-Late 20th Century):** The market has been predominantly characterized by the prevalence of

internal combustion engine (ICE) vehicles, with advancements in performance, range, and infrastructure developing over decades.

Development of Electric Vehicles (EVs) 1890-Present

Early EV Popularity (1890-1920s):

Innovation: The late 1800s saw significant advances in battery technology, particularly the invention of lighter, more efficient lead-acid batteries. By innovators like Camille Faure.

A Golden Age: EVs were popular for their quietness, cleanliness, and ease of operation, accounting for a significant portion of all cars around 1900.

Fig.1.2: Old Carrier Vehicle

Decline: The mass production and affordability of the Model T, along with limited EV range and charging infrastructure outside of urban centers, led to a decrease in EV popularity by the 1920s.

Revival and Challenges (1970s-1990s):

First Revival: The oil crises of the 1970s were pivotal events that reshaped global economic and energy policies, renewing scholarly and industrial focus on alternative fuel vehicles, including EVs.

Stagnation: However, the limited performance, short driving range, and high costs of late-20th-century EVs hampered widespread adoption.

Environmental Regulations: New environmental regulations in the 1990s boosted development efforts, leading to improved technologies and the emergence of vehicles like GM's EV1.

Modern Resurgence (2000s-Present):

Mass-Produced Hybrids: The Toyota Prius, the first mass-produced hybrid vehicle, was launched in 1997, paving the way for electric-hybrid technology.

Battery Advancements: Battery costs have dropped significantly, and lithium-ion battery technology has enabled longer ranges and better performance.

Commercial Availability: The 2010s saw the launch of mass-produced is commercially available. The expansion of the electric vehicle (EV) market, ignited by mass-market vehicles such as the Nissan LEAF, is fundamentally intertwined with the evolution of a national charging infrastructure. The progress of this infrastructure, which includes everything from standard residential Level 2 chargers to high-speed public DC fast chargers, is the key to accelerating EV adoption. Rising environmental awareness, reduced lifetime emissions, and the expanding range of options in today's market have driven the growing interest and sales of electric vehicles. To support this transition, central authorities introduced financial incentives, including tax credits, subsidies, and rebates, to make EVs more affordable and attractive to consumers. During this period, several influential players emerged in the EV market, including Tesla, Nissan, BYD, and Rivian, each fostering unprecedented innovation, commercialization, and a surge in EV popularity. Simultaneously, an expansion of public charging infrastructure—ranging from fast chargers to Tesla's Supercharger network—played a pivotal role in easing range anxiety and making EV adoption more feasible on a global scale.

Another transformative development has been the seamless merging of electric vehicles with a smart grid system. The shift to viewing electric vehicles as more than just transport, but as a key element of energy storage. This vision supports dynamic energy balancing, allowing EVs to interact with the grid by using V2G applications to boost the resilience and efficiency of our energy ecosystem.

1.2 Smart Grid: A Sustainable Solution Towards Stability and Efficiency

A smart grid represents an advanced electricity network that leverages digital communication, sensors, and automation to optimize the generation, distribution, and consumption of power. In relation to electric vehicles (EVs), it enables an intelligent, bidirectional exchange of energy and information between EVs and the power grid. EVs thus evolve from being mere electricity consumers to functioning as distributed energy storage systems that contribute to grid stability and operational efficiency.

Through smart charging, vehicles can automatically draw power during off-peak hours—when electricity is more affordable, and demand is minimal—reducing grid stress and user costs. Moreover, Vehicle-to-Grid (V2G) technology allows EVs to feed stored energy back into the grid during high-demand periods, effectively serving as mobile energy reservoirs.

Smart grids also enable demand response mechanisms, allowing charging to be delayed or adjusted based on real-time grid conditions to help maintain balance and reliability. This interconnected framework becomes increasingly important with the expansion of renewable energy sources like solar and wind, which are naturally intermittent. EVs can store surplus power during high generation periods and release it when supply dips, promoting grid equilibrium and improving the overall integration efficiency of renewables. Examples from around the world include Nissan's V2G trials in Japan, Tesla's solar-vehicle-grid ecosystem, and PG&E's incentive-based EV-grid interaction programs in California.

Underlying technologies such as smart meters, bi-directional chargers, IoT-enabled charging stations, and AI-driven load forecasting make this seamless interaction possible. As EV adoption rises, their role within the smart grid is expected to grow, enabling future innovations such as peer-to-peer energy trading, home-grid backup systems, and autonomous energy balancing. Ultimately, integrating EVs into the smart grid not only enhances grid resilience and energy security but also lays the foundation for a more sustainable and decentralized energy system and decentralized energy ecosystem.

Merits & Demerits of Electric Vehicles GTUexam-3170923-SUM2025

Demerits: A significant disadvantage is the high upfront purchase price of EVs, which can be a barrier for many buyers compared to conventional vehicles. Another challenge is the limited charging infrastructure and the longer time it takes to charge an EV battery compared to filling a gas tank, leading to "range anxiety" for drivers on longer journeys

Merits: Electric vehicles (EVs) produce zero tailpipe emissions, significantly improving air quality and reducing the carbon footprint of transportation. They also offer lower running costs due to cheaper electricity compared to fuel and significantly reduced maintenance needs because of fewer moving parts and no oil changes.

1.3 Battery Electric Vehicles GTUexam-3170923-SUM2025

Battery Electric Vehicles (BEVs) operate using an electric battery for energy storage, an electric motor for propulsion, and a controller to regulate power flow. The battery is recharged from the mains electricity supply via a plug, using a battery charging unit that may be onboard or installed at the charging point. The controller plays a vital role by managing the power delivered to the motor, thus controlling the vehicle's speed in both forward and reverse directions. This setup is referred to as a 2-quadrant controller, which handles propulsion in two directions. When the controller also enables regenerative braking—allowing energy recovery while decelerating in both directions—it becomes a 4-quadrant controller, enhancing efficiency and performance.

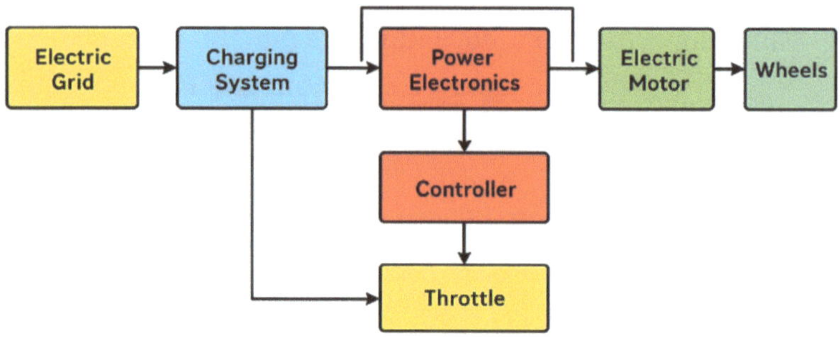

Fig.1.3: Battery Electric Vehicle

Battery electric vehicles come in a variety of sizes and configurations to suit different applications. At the most basic level, there are small electric bicycles, tricycles, and commuter vehicles, which are lightweight and ideal for short-distance travel. On the larger end of the spectrum, BEVs include full-sized electric vehicles such as electric cars, delivery trucks, and buses, all of which contribute to reducing emissions and reliance on fossil fuels. The adoption of BEVs continues to grow due to technological advancements, market availability, and environmental benefits.

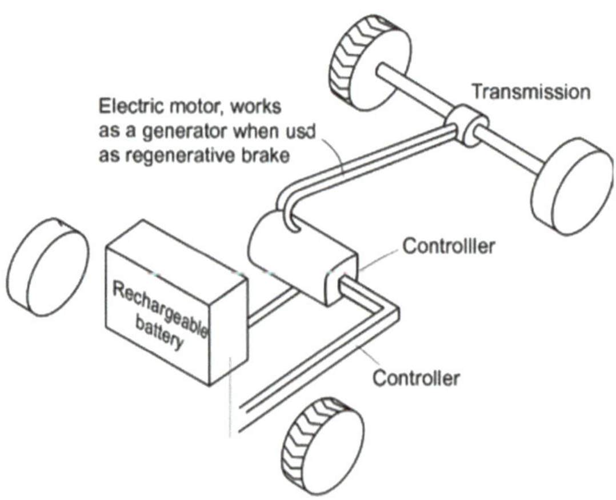

Fig.1.4: Battery Placement-BEV

1.4 EV adoption

EV adoption leads to improvements in range, cost reduction, and the achievement of global sustainability goals. The widespread adoption of electric vehicles (EVs) has played a pivotal role in advancing multiple aspects of automotive and energy technology. As the EV market continues to grow, it drives significant improvements in battery range, allowing modern electric vehicles to travel longer distances on a single charge—addressing one of the early concerns of limited range. Additionally, economies of scale, technological innovation, and competitive manufacturing have contributed to a steady reduction in the overall cost of EVs, making them increasingly affordable for the average consumer. More importantly, the transition toward EVs aligns directly with global sustainability goals, such as reducing greenhouse gas emissions, decreasing reliance on fossil fuels, and improving urban air quality. Thus, EV adoption is not only transforming transportation but also supporting the broader shift toward a cleaner, more sustainable future.

1.5 Motion and Dynamic Equations of Electric Vehicles

Electric vehicles obey Newton's laws of motion. Their performance is governed by the balance between tractive forces that propel the vehicle and resistive forces that oppose motion.

1.5.1 Fundamentals of Vehicle Motion

The longitudinal motion of an electric vehicle is determined by evaluating the forces acting along its direction of travel. Vehicle dynamics are primarily governed by classical mechanics, with Newton's second law providing the foundation for motion analysis. This law states that a vehicle experiences acceleration only when the resultant force acting on it is non-zero. In practical driving conditions, multiple forces act on the vehicle simultaneously, and their combined effect determines its motion. The traction force generated by the propulsion system enables forward movement by counteracting opposing forces such as gravitational components, aerodynamic drag, and rolling resistance from the tires. Consequently, the magnitude of vehicle acceleration is governed by the balance between the driving force and these resistive forces.

- Power delivered by the propulsion unit
- Road conditions
- Aerodynamics of the vehicle
- Composite mass of the vehicle

Newton's laws, particularly his second law, F=ma, are fundamental to understanding the motion and dynamics of electric vehicles. These laws relate the forces acting on a vehicle to its acceleration. In simple terms, the net force on the vehicle equals its mass times its acceleration. This force is influenced by the electric motor's thrust, gravity, rolling resistance, and aerodynamic drag. In this description, the mathematical framework required for the analysis of vehicle mechanics based on Newton's second law of motion is presented. The following topics are highlighted in the description

- General description of vehicle movement
- Vehicle resistance
- Dynamic equation
- Tire Ground Adhesion and maximum tractive effort

1.5.2 General Description of Vehicle Movement

The vehicle motion can be completely determined by analyzing the forces acting on it in the direction of motion. The forces acting on a vehicle, moving up a grade, are shown in Figure. The tractive force Ft in the contact area between the tires of the driven wheels and the road surface propels the vehicle forward. The tractive force Ft, is produced by the power plant and transferred to the driving wheels via the transmission and the final drive. When the vehicle moves, it encounters a resistive force that tries to retard its motion. The resistive forces are

- Rolling resistance
- Aerodynamic drag
- Uphill resistance

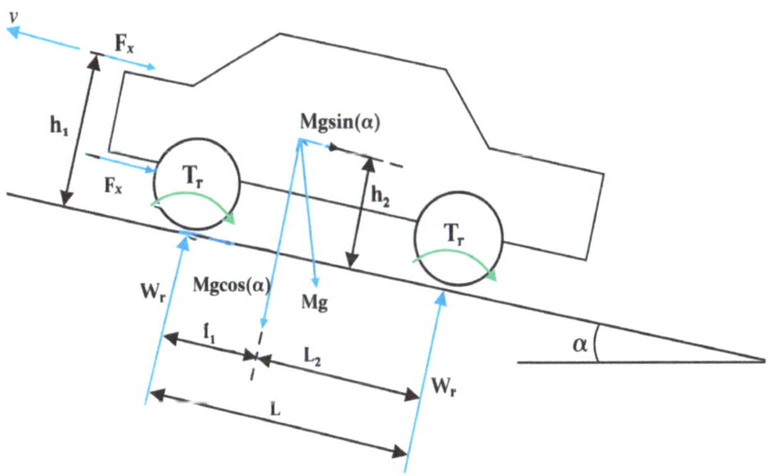

Fig.1.5: Force Placement

1.5.3 Rolling Resistance GTUexam-3170923-SUM2025

The rolling resistance of tires on hard surfaces arises primarily due to hysteresis in the tire material. This hysteresis is a result of the energy losses occurring during the continuous deformation and recovery of the tire as it rolls over a surface.

Forces acting on an electric vehicle:

- **Tractive Force (Ft):** This is the force generated by the electric motor, propelling the vehicle forward.
- **Rolling Resistance (Fr):** This force opposes motion due to the tires deforming as they roll.
- **Aerodynamic Drag (Fd):** This force opposes motion due to air resistance, increasing with speed.

- **Grade Resistance (Fg):** This force opposes motion when the vehicle is on an incline.

$$a = \frac{F_t - (F_r + F_d + F_g)}{m}$$

Where:

- a = vehicle acceleration
- m = vehicle mass

The vehicle motion can be completely determined by analysing the forces acting on it in the direction of motion. The forces acting on a vehicle, moving up a grade, are shown in Figure. The tractive force (Ft) in the contact area between the tires of the driven wheels and the road surface propels the vehicle forward. The tractive force (Ft) is produced by the power plant and transferred to the driving wheels via the transmission and the final drive. When the vehicle moves, it encounters a resistive force that tries to retard its motion.

The following are the resistive forces:

Rolling resistance arises mainly due to hysteresis losses in the tire material, caused by repeated deformation and recovery.

Major Forces Acting on an EV:

- Tractive Force: From motor torque and gear ratio
- Rolling Resistance: Depends on tire pressure, road surface, and speed
- Aerodynamic Drag: Proportional to velocity squared, influenced by frontal area and drag coefficient
- Grade Resistance: Proportional to vehicle weight and road slope angle

Other Dynamic Factors:

- Rotational Inertia: Affects how the vehicle responds to applied torque

- Suspension: Governs load transfer and handling stability
- Tire–Road Interaction: Determines traction and adhesion limits
- Energy Consumption: Estimated from force balance and driving cycle

Energy Consumption:

The dynamic equations can be used to estimate energy consumption based on speed, acceleration, and other factors. In the Figure, a tire at a standstill is illustrated. A vertical force P is applied at the centre of the tire. The pressure in the contact area between the tire and the ground is distributed symmetrically about the tire's centreline. Consequently, the resulting reaction force p r is aligned directly below the applied force P, and no net torque is generated.

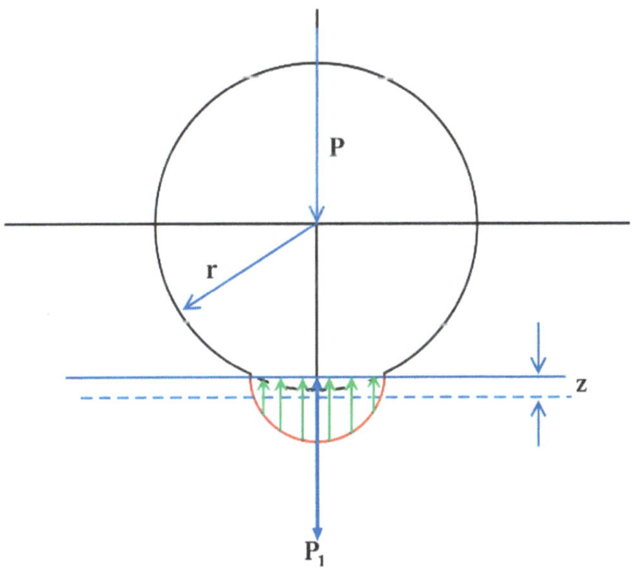

Fig.1.6: Pressure Distribution

In summary, by understanding the forces acting on an electric vehicle and applying Newton's laws, engineers can model and predict its motion and performance, which is crucial for vehicle design and control.

1.6 Forces acting on an electric vehicle GTUexam-3170923

Tractive force (Ft) is the fundamental driving force that enables a vehicle to move forward, as it is generated by the electric motor and transmitted to the wheels through the drivetrain. This force must be sufficient to overcome all opposing forces acting on the vehicle to achieve motion or maintain a desired speed. Opposing this motion is rolling resistance (Fr), which arises primarily due to the deformation of tires and road surface contact as the wheels roll; although relatively small, it is always present and depends on factors such as tire pressure, vehicle weight, and road conditions. Another major opposing force is aerodynamic drag (Fd), caused by air resistance acting against the vehicle's motion, and this force increases rapidly with speed, following a quadratic relationship with velocity, making it especially significant at high speeds such as highway driving. Additionally, grade resistance (Fg) comes into effect when a vehicle travels on an inclined surface, where a component of the vehicle's weight acts opposite to the direction of motion, increasing the demand on the motor during uphill travel and assisting motion during downhill movement. Conceptually, vehicle motion can be visualized as Ft → [Vehicle] ← (Fr + Fd + Fg), where forward motion occurs only when Ft exceeds the sum of resistive forces. For example, if an electric vehicle produces a tractive force of 3000 N while experiencing rolling resistance of 400 N, aerodynamic drag of 1100 N, and grade resistance of 900 N on an incline, the net force becomes 600 N, allowing the vehicle to accelerate forward. Therefore, understanding the balance between tractive force and resistive forces is crucial in electric vehicle design, performance analysis, and energy efficiency optimization, as it directly influences acceleration capability, maximum speed, and overall power consumption in real-world driving conditions.

Simplified Dynamic Equation GTUexam-3170923-SUM2025

The basic equation for the vehicle's longitudinal acceleration (a) can be expressed as:

$$a = \frac{(Ft - Fr - Fd - Fg)}{m}$$

Where:

- a is the acceleration.
- Ft is the tractive force.
- Fr is the rolling resistance force.
- Fd is the aerodynamic drag force.
- Fg is the grade resistance force.
- m is the mass of the vehicle.

Further Considerations:

Tractive force (Ft) represents the driving effort produced at the wheels and is primarily governed by the torque generated by the electric motor and the selected gear ratio, which together determine how effectively power is transmitted to the road. A higher motor torque or gear reduction results in greater tractive force, improving acceleration and load-carrying capability. Opposing this motion is rolling resistance (Fr), a force that originates from tire deformation and friction at the tire–road interface. Its magnitude is influenced by tire inflation pressure, road texture, vehicle mass, and to a lesser extent, operating speed; underinflated tires or rough road surfaces significantly increase this resistance. Aerodynamic drag (Fd) becomes dominant at higher vehicle speeds and arises from air flowing around the vehicle body. This force increases with the square of the vehicle's velocity and is strongly affected by the frontal area and drag coefficient, making streamlined vehicle design critical for energy efficiency. Another important opposing force is grade resistance (Fg), which occurs when a vehicle travels on an inclined road. This force depends on the vehicle's total weight and the sine of the road angle, meaning steeper slopes demand substantially higher driving force. Together, these forces define the overall performance and efficiency of a vehicle, and the tractive force must continuously balance or exceed the combined resistances to ensure smooth and controlled motion.

Dynamic Model:

A more complete dynamic model would also consider factors like:

- *Rotational Inertia:*
  The vehicle's rotational inertia (how easily it rotates) affects its response to forces.
- *Suspension:*
  The suspension system influences how forces are transmitted to the chassis and how the vehicle handles.
- *Tire-Road Interaction:*
  The interaction between the tires and the road surface is complex and affects traction and stability.
- *Energy Consumption:*
  The dynamic equations can be used to estimate energy consumption based on speed, acceleration, and other factors.

Example: Compute aerodynamic drag, rolling resistance & gradient force for the following vehicle 07 assuming ρ=1.2 kg/m3 & θ = 8°. Gross vehicle weight = 1500 kg, CD=0.3, Area = 2.5 m2, μ=0.015, v = 80 kmph, Tyre radius = 0.3 m. chapt3 (ii) Also compute total traction force assuming pick-up from 0 to 50 kmph in 20 sec, with linear acceleration & zero slope. Uphill forces at 8° (v = 80 km/h)

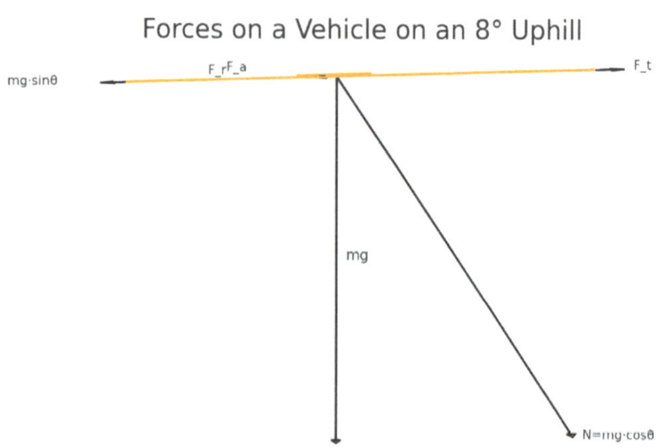

Fig.1.7: Force Placement

Total traction during pickup 0→50 km/h in 20 s (flat road) Given Data:

$\rho = 1.2\ \text{kg/m}^3$
$\theta = 8°$
$m = 1500\ \text{kg}$
$Cd = 0.3$
$A = 2.5\ \text{m}^2$
$\mu = 0.015$
$v = 80\ \text{km/h} = 22.22\ \text{m/s}$
$g = 9.81\ \text{m/s}^2$

Aerodynamic Drag Force

$F_a = ½ \times \rho \times A \times Cd \times v^2$
$F_a = 0.5 \times 1.2 \times 2.5 \times 0.3 \times (22.22)^2$
$F_a \approx 222.22\ \text{N}$

Rolling Resistance Force

$F_r = m \times g \times \cos\theta \times \mu$
$F_r = 1500 \times 9.81 \times \cos 8° \times 0.015$
$F_r \approx 218.58\ \text{N}$

Gradient Force

$F_g = m \times g \times \sin\theta$
$F_g = 1500 \times 9.81 \times \sin 8°$
$F_g \approx 2047.93\ \text{N}$

Total Tractive Force (*No Acceleration*)

$F_t = F_a + F_a + F_g$
$F_t \approx 222.22 + 218.58 + 2047.93 \approx 2488.73\ \text{N}$

Now compute the total traction force for pick-up from 0 to 50 km/h in 20 seconds on a flat road.

Given Data:

$v_f = 50\ \text{km/h} = 13.89\ \text{m/s}$
$t = 20\ \text{s}$

Acceleration

$a = (vf - vi) / t$
$a = 13.89 / 20 \approx 0.6944 \text{ m/s}^2$

Inertial Force

$F_i = m \times a$
$F_i = 1500 \times 0.6944 \approx 1041.67 \text{ N}$

Aerodynamic Drag (50 km/h)

$F_a = 0.5 \times 1.2 \times 2.5 \times 0.3 \times (13.89)^2$
$F_a \approx 86.81 \text{ N}$

Rolling Resistance (Flat Road)

$F_r = m \times g \times \mu$
$F_r = 1500 \times 9.81 \times 0.015 \approx 220.73 \text{ N}$

Total Traction Force

$F_t = F_a + F_r + F_i$
$F_t \approx 86.81 + 220.73 + 1041.67 \approx 1349.21 \text{ N}$

Final Answer: Uphill (8°) at 80 km/h; Total Tractive Force = 2488.73 N

Pick-up (0 → 50 km/h in 20 s, flat); Total Traction Force = 1349.21 N

1.6.1 Effect of Hysteresis on Rolling

When the tire is in motion, the viscoelastic nature of rubber causes energy loss due to internal friction during loading and unloading cycles. This hysteresis leads to an asymmetric distribution of the ground reaction forces.

This behavior is shown in Fig, which plots the applied force PP against deformation zz. During the loading phase, the force increases more rapidly than in the unloading phase for the same amount of deformation, leading to a force hysteresis loop. This asymmetry causes the resultant ground reaction force to shift slightly ahead of the vertical centerline, creating a moment that opposes the tire's rotation — perceived as rolling resistance. Motion can be completely determined by analyzing the forces acting on it. This behavior is shown in Figure, which plots the applied force PP against deformation zz. During the loading phase, the force increases more rapidly than in the unloading phase for the same amount of deformation, leading to a force hysteresis loop. This asymmetry causes the resultant ground reaction force to shift slightly ahead of the vertical centerline, creating a moment that opposes the tire's rotation- perceived as rolling resistance. Motion can be completely determined by analyzing the forces acting on it.

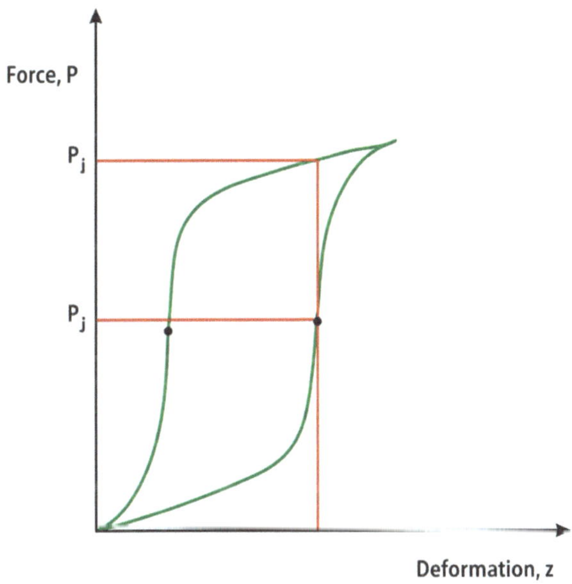

Fig 1.8: Pressure Distribution

1.7 Types of EV

1.7.1 Battery Electric Vehicles (BEVs)

- **How they work:** Powered entirely by an electric motor and a large battery pack that is recharged by plugging into an external power source.
- **Characteristics:** Zero tailpipe emissions and a long driving range, typically, 150-400 miles.

- **Examples:** Tata Nexon EV, MG Comet EV, and other all-electric cars available in India.

Fig 1.9: EV

1.7.2 Plug-in Hybrid Electric Vehicles (PHEVs)

- **How they work:** A combination of an electric motor, a larger battery, and a gasoline engine. The battery can be charged externally and allows for a moderate distance on electricity alone.
- **Characteristics:** Offers flexibility with the option to use only electric power for shorter trips or the gasoline engine for longer journeys.

Fig 1.10: PHEV

1.7.3 Hybrid Electric Vehicles (HEVs) GTUexam-**3170923-WIN2024**

- **How they work:** Use both a gasoline engine and an electric motor, but the electric battery is smaller and recharged by the engine or during braking, rather than being plugged in.
- **Characteristics:** More fuel-efficient than conventional gasoline cars, but cannot be driven solely on electricity for extended periods.

Fig 1.11: HEV

1.7.4 Fuel Cell Electric Vehicles (FCEVs)

- **How they work:** Use a fuel cell to convert hydrogen gas into electricity, which then powers the electric motor.

Fig 1.12: FCEV

- **Characteristics:** Zero tailpipe emissions, quick refueling times (similar to gasoline cars), and a long driving range.
- **Examples:** Toyota Mirai and Hyundai Nexo are available models.

1.7.5 Solar Electric Vehicles (SEVs)

- **How they work:** Solar panels are integrated into the vehicle to convert sunlight into electricity to charge the battery, supplementing the main power source.
- **Characteristics:** Uses free, clean energy from the sun to extend range, but requires sunlight for optimal performance.

The following table shows the classification of EV

Table 1.1: Classification of EV

Type	Power Source	Battery Charging	Driving Range	Fuel/ Energy Used	Examples
BEV (Battery Electric Vehicle)	Electric motor only	Plug-in external charging	150–400 miles	Electricity	Tata Nexon EV, MG Comet EV
PHEV (Plug-in Hybrid EV)	Electric motor + Gasoline engine	Plug-in charging + gasoline refueling	Moderate (electric) + Long (hybrid)	Electricity + Petrol/ Diesel	Toyota Prius Plug-in
HEV (Hybrid EV)	Gasoline engine + Electric motor	Self-charged (engine + regenerative braking)	Limited electric + Long (hybrid)	Petrol/ Diesel	Toyota Camry Hybrid

FCEV (Fuel Cell EV)	Electric motor powered by hydrogen fuel cell	Refueling with hydrogen gas	Long (comparable to petrol cars)	Hydrogen	Toyota Mirai, Hyundai Nexo
SEV (Solar EV)	Electric motor + Solar panels	Sunlight + plug-in charging (if available)	Limited, depends on sunlight	Solar energy + Electricity	Aptera SEV (concept), Lightyear One

1.7.6 Advantages of AC Motors over DC Motors GTUexam-3170923-WIN2025

In electric vehicle applications, AC motors such as Induction Motors and Permanent Magnet Synchronous Motors (PMSM) are increasingly favored over DC motors due to their superior performance and efficiency. One of the primary advantages of AC motors is their higher efficiency, especially at varying speeds, which helps improve the overall range and energy utilization of EVs. Unlike DC motors, AC motors do not require brushes or commutators, eliminating mechanical wear and reducing maintenance requirements significantly. This not only enhances durability but also lowers long-term operational costs. Another major benefit is better speed control. AC motors, with the help of Variable Frequency Drives (VFDs) or inverters, provide a wide speed range and precise control, which is essential for smooth acceleration and performance optimization in electric vehicles. Additionally, AC motors offer higher power density, meaning they can produce greater torque for a given size and weight, allowing for compact and lightweight designs that are ideal for EV architecture. AC motors also have inherent regenerative braking capability, allowing them to switch to generator mode during braking, recovering energy and feeding it back to the battery, thus improving vehicle efficiency. Their design with fewer mechanical components results in superior durability and reliability, making them suitable for long-term operation under harsh driving conditions. Moreover, AC motors perform exceptionally well at high speeds, which is crucial for highway driving and high-performance electric vehicles. When compared to DC motors, AC motors stand out in terms of maintenance, efficiency, speed control, compactness, regenerative braking ease, and overall durability. DC motors, on the other hand, tend to have higher maintenance needs due to brush wear, provide moderate efficiency, limited speed control, and are bulkier in design. Therefore, AC motors have become the preferred choice for modern electric and hybrid vehicles.

Comparison Chart: AC vs DC Motors for EVs GTUexam-3170923-WIN202

Parameter	**AC Motor (Induction/ PMSM)**	**DC Motor**
Maintenance	Low (No brushes)	High (Brush wear)
Efficiency	High at all speeds	Moderate
Speed Control	Precise (via inverter)	Limited
Weight & Size	Compact	Bulkier
Regenerative Braking	Easy to implement	Complex
Durability	High	Lower

Remembrance: Higher Efficiency-AC motors like Induction Motors and Permanent Magnet Synchronous Motors operate at higher efficiency levels compared to DC motors, especially at varying speeds. Lower Maintenance-AC motors do not use brushes or commutators (as in DC motors), eliminating mechanical wear and reducing maintenance requirements. Better Speed Control-AC motors offer a wide speed range and precise control using Variable Frequency Drives (VFDs) or inverters, which is essential for EV performance. Higher Power Density-AC motors provide better torque-to-weight ratio, making them more compact and lighter for the same output. Regenerative

Braking Capability-AC motors can easily switch to generator mode during braking, recovering energy and improving vehicle efficiency. Durability & Reliability-With fewer mechanical parts and a robust design, AC motors offer longer life and reliable performance under harsh driving conditions. Better High-Speed Performance- AC motors perform efficiently at higher speeds, which is critical for highway driving and performance EVs.

Multiple Choice Questions

Question

1.1 What is the total tractive effort required to move an electric vehicle?

A) Only aerodynamic drag
B) Sum of all resistive and inertial forces
C) Rolling resistance only
D) Engine torque alone

1.2. Which force acts opposite to the vehicle's motion due to tire-road contact?

A) Inertial force
B) Rolling resistance
C) Gradient force
D) Normal force

1.3. What does the term represent?

$$F_a = \frac{1}{2}\rho C_d A v^2$$

A) Engine force
B) Rolling resistance
C) Aerodynamic drag
D) Gradient force

1.4. In static conditions, the net force acting on the electric vehicle is:

A) Equal to engine power
B) Zero
C) Equal to the gradient force
D) Equal to rolling resistance

1.5. If a 1500 kg vehicle accelerates at 3 m/s², what is the inertial force?

A) 4500 N

B) 500 N

C) 1500 N

D) 750 N

1.6. Which factor does not affect the aerodynamic drag on a vehicle?

A) Vehicle speed

B) Drag coefficient

C) Mass of vehicle

D) Frontal area

1.7. The normal reaction force on a sloped road is given by:

A) $N = mg\sin(\theta)$

B) $N = mg\cos(\theta)$

C) $N = mg\tan(\theta)$

D) $N = mg$

1.8. What is the unit of power required by the EV motor?

A) Newton

B) Joule

C) Watt

D) Meter/second

1.9. Which force is primarily responsible for slowing the vehicle when going uphill?

A) Rolling resistance

B) Aerodynamic drag

C) Gradient resistance

D) Tractive effort

2.0 If vehicle's speed is doubled, aerodynamic drag:

A) Doubles

B) Increases 4 times

C) Increases 8 times

D) Remains constant

Answers:

1.1 B) Sum of all resistive and inertial forces

Explanation: Total tractive force = Fr + Fg + Fa + FiFr + Fg + Fa + Fi

1.2 B) Rolling resistance

Explanation: Rolling resistance is caused by deformation at the tire-road interface.

1.3 C) Aerodynamic drag

Explanation: Drag increases with square of speed and depends on vehicle shape.

1.4 B) Zero

Explanation: In static equilibrium, all forces cancel out.

1.5 A) 4500 N

Explanation: F = ma = 1500 × 3 = 4500 N

1.6 C) Mass of vehicle

Explanation: Drag is independent of vehicle mass

1.7 B) $N = mg \cos(\theta)$

Explanation: Normal force is the vertical component of weight on an incline.

1.8 C) Watt

Explanation: P = T · vP gives power in watts.

1.9 C) Gradient resistance

Explanation: Fg = mg * sin(θ), acts opposite to motion uphill.

2.0 B) Increases 4 times

Explanation: Drag is proportional to v^2 v^2

India-GTU Questions

Explain the different types of forces acting on the vehicle due to air friction.	chapt1pg11,12GTU exam-3170923-SUM2025
Derive equations & explain different types of forces acting on a vehicle going uphill with neat diagram.	chapt1pg11,12GTU exam-3170923-SUM2025
Compare hybrid electric vehicles with conventional vehicles.	chapt2GTUexam-3170923-SUM2025
Write two merits and two demerits of electric vehicles.	chapt1pg7GTU exam-3170923-SUM2025
Explain historical development of automobile and development of interest and activity in the EV from 1890 to the present day.	chapt1pg3GTU exam-3170923-SUM2025
Derive equations & explain different types of forces acting on vehicle going uphill with neat diagram.	chapt1pg11,12,14 GTUexam-3170923-WIN2025
Write two merits and two demerits of electric vehicles.	chapt1pg7 GTUexam-3170923-WIN2025
What is the advantage of an AC motor over a DC motor for EV applications?	chapt1pg23 GTUexam-3170923-WIN2025

Chapter 02

Introduction to Hybrid Electric Vehicles

2 Introduction to Hybrid Electric Vehicles

2.1 Social and Environmental Importance of Hybrid Electric Vehicles (HEVs)

Hybrid Electric Vehicles (HEVs) represent a significant step forward in the pursuit of sustainable and responsible mobility. By integrating an internal combustion engine with an electric propulsion system, HEVs offer a compelling solution to many of the environmental and societal challenges posed by traditional fossil-fuel-based vehicles. One of the foremost benefits of HEVs lies in their environmental impact. Compared to conventional internal combustion engine (ICE) vehicles, HEVs significantly reduce greenhouse gas emissions, nitrogen oxides (NO_x), and particulate matter (PM). These reductions not only contribute to cleaner air but also play a crucial role in mitigating climate change.

From the perspective of energy security, HEVs help decrease dependence on imported fossil fuels. Their ability to operate partially on electricity allows for greater integration with renewable energy sources, supporting national and global objectives for sustainable energy use. Moreover, the quiet operation of electric motors in urban settings helps reduce noise pollution, creating more peaceful and livable cities. In terms of fuel economy, HEVs provide higher mileage compared to traditional ICE vehicles. Technologies such as regenerative braking enable the vehicle to recover and reuse energy, which would otherwise be lost as heat, thereby improving overall energy efficiency and reducing fuel consumption.

HEVs are also essential in advancing climate action. They support international commitments such as the Paris Agreement (COP21), and help achieve national climate goals, including those laid out in India's FAME-II (Faster Adoption and Manufacturing of Electric Vehicles) scheme. For instance, the Toyota Prius, a well-known hybrid model, emits approximately 50% less carbon dioxide (CO_2) than a comparable non-hybrid petrol vehicle, showcasing the tangible benefits of hybrid technology.

The social benefits of HEVs extend to public health as well. Reduced vehicular emissions lead to improved air quality, which in turn decreases respiratory illnesses and enhances the overall well-being of urban populations. Additionally, HEVs act as a transitional technology by fostering consumer awareness and acceptance of electric mobility, paving the way for future advancements in battery electric vehicles (BEVs) and hydrogen fuel cell vehicles.

In the Indian context, HEVs align well with environmental and economic priorities. As India continues to implement stricter emission standards like Bharat Stage VI (BS-VI), HEVs provide an effective means of compliance. Furthermore, by lowering the nation's dependence on crude oil imports, HEVs contribute to economic resilience and energy sovereignty. With support from government policies and incentives such as FAME-II, the adoption of HEVs is expected to grow, making them a pivotal part of India's mobility transition.

In summary, HEVs serve as a bridge between conventional ICE vehicles and full electric vehicles, offering a balanced mix of environmental responsibility, energy efficiency, and practical usability. Their adoption not only reflects technological advancement but also demonstrates a commitment to sustainability, public health, and climate resilience. Example: Toyota Prius emits ~50% less CO_2 than a similar ICE vehicle.

Key points: Environmental, Energy Security, Noise Pollution, Fuel Economy, Climate Goals

Linked poll: Which segment is likely to lead EV growth in the next 5 years?

Battery Electric Vehicles-BEVs	67%
Plug-in Hybrid Vehicles-PHEVs	0%
Fuel Cell Electric Vehicles	0%
Two & three-wheeler EVs	33%

2.1.1 Need and importance of electric and hybrid electric vehicles GTUexam-3170923-WIN2024

Electric and hybrid vehicles are needed and important because they significantly reduce air pollution and greenhouse gas emissions, lower fuel costs, and decrease dependence on fossil fuels, contributing to a healthier environment and a more sustainable, resilient transportation system. EVs offer zero tailpipe emissions and lower maintenance, while HEVs provide better fuel efficiency by combining gas and electric power, making both appealing choices for reducing carbon footprints and running costs.

Environmental Benefits-Reduced Air Pollution and Greenhouse Gases:

EVs produce zero tailpipe emissions, and HEVs produce substantially fewer emissions than conventional cars, leading to cleaner air and a healthier planet.

Lower Carbon Footprint-By relying on electricity and more efficient fuel use, these vehicles help reduce the overall carbon footprint associated with transportation.

Economic Benefits-Lower Running Costs:

Using electricity instead of petrol or diesel, combined with more efficient energy use, makes charging and operating these vehicles cheaper.

Lower Maintenance Costs-EVs and HEVs have fewer moving parts than traditional vehicles, which reduces wear and tear, leading to lower maintenance needs and costs.

Reduced Dependence on Fossil Fuels-Shifting to electric power reduces reliance on petroleum, making transportation more resilient and less vulnerable to volatile fuel prices.

Driving and Technology Benefits

- Enhanced Driving Experience:

Hybrids offer improved driving experience due to advanced technology, and electric vehicles can be quieter and smoother to drive.

- Increased Durability:

The combination of engines in hybrids can lead to less wear on components, extending the vehicle's lifespan, according to SEAT.

- Government Incentives:

Many governments offer financial incentives, such as tax credits, to encourage the adoption of electric and hybrid vehicles.

Societal Benefits

- Improved Public Health: Cleaner air from lower emissions leads to better public health outcomes.
- Resilient Transportation System: Diversifying energy sources for transportation creates a more robust and reliable system.

Key Qualities of Electric and Hybrid Vehicles (EVs & HEVs)

Electric and Hybrid Vehicles must integrate advanced design and performance characteristics to ensure efficiency, reliability, and sustainability.

1. High Energy Efficiency

- EVs and HEVs must optimize the conversion of stored energy into mechanical motion, reducing losses in batteries, power electronics, and motors.

- Ensures longer driving range and reduced operating costs.

2. Lightweight and Compact Design

- Lightweight chassis and components improve power-to-weight ratio, acceleration, and efficiency.
- Compact powertrains free up cabin and cargo space.

3. Advanced Energy Storage Systems

- High-performance batteries, fuel cells, or supercapacitors with high energy density, fast charging, and long lifecycle.

4. Regenerative Braking Capability

- Recovers kinetic energy during braking and converts it into electrical energy to recharge the battery, improving overall efficiency.

5. Low Emissions and Environmental Impact

- EVs have zero tailpipe emissions, while HEVs minimize emissions using hybrid drive modes and efficient combustion strategies.

6. High Reliability and Durability

Components must withstand thermal stresses, frequent load changes, and harsh environmental conditions without failure. Low Noise, Vibration, and Harshness (NV Silent operation enhances driving comfort and supports urban noise reduction.

Advanced Control Systems-Intelligent energy management, torque vectoring, and drive control algorithms for optimal performance under varying driving conditions.

Fast Charging and Long Range-Ability to support high-voltage fast charging while maintaining extended range to compete with conventional vehicles.

Safety and Thermal Management-Advanced safety systems for battery protection, crash safety, and efficient cooling to maintain performance and prevent overheating.

2.1.2 Regenerative Braking System GTUexam-3170923-SUM2025

An electronically controlled regenerative braking system functions in coordination with an Anti-lock Braking System (ABS), not as an ABS itself, to ensure vehicle stability and energy efficiency during braking. The system integrates both regenerative and hydraulic friction brakes, using a sophisticated controller to manage the vehicle's braking torque. During heavy or emergency braking, the controller monitors wheel speed to prevent lock-up, adjusting the distribution between regenerative and friction braking to maintain optimal wheel slip and steering control, similar to a conventional ABS, while also recovering kinetic energy for the battery.

How it Works

1. ABS Functionality First:

At the core of the system is the ABS, which constantly monitors wheel speeds to detect impending wheel lock-up.

2. Brake Torque Allocation:

A sophisticated controller decides how much braking force to apply. It prioritizes using the electric motor (acting as a generator) for regenerative braking to capture energy.

3. Coordinated Action:

- **Normal Braking:** The system primarily uses regenerative braking to recover energy, slowing the vehicle.
- **Heavy or Emergency Braking:** If the ABS detects that the wheels are about to lock up, it will modulate the amount of regenerative braking applied. It can also blend in hydraulic (friction) brakes as needed to achieve the desired stopping power while keeping the wheels rotating, preventing skidding.

4. Energy Recovery:

When regenerative braking is used, the electric motor converts the vehicle's kinetic energy into electrical energy, which is then stored in the vehicle's battery.

5. Stability and Control:

By ensuring the wheels don't lock up, the system maintains directional stability and the driver's ability to steer during hard braking, even on slippery surfaces.

Key Aspects

- **Blended Systems:**

Rather than relying solely on regenerative braking, the system blends it with conventional friction brakes.

- **Smart Controllers:**

Advanced control algorithms, such as fuzzy logic or model predictive control, are used to manage the complex coordination between the regenerative and hydraulic braking systems.

- **Energy Efficiency:**

The primary goal is to recover as much kinetic energy as possible for storage in the battery, which increases the vehicle's driving range.

- **Safety:**

By functioning like a conventional ABS, the system prevents accidents by ensuring the vehicle remains stable and controllable under all braking conditions

2.1.3 Comparison of Hybrid Electric Vehicles (HEVs) and Conventional Vehicles (CVs) GTUexam-3170923-SUM2025

Hybrid Electric Vehicles (HEVs) and Conventional Vehicles (CVs) are two distinct categories of automotive technology, differing significantly in their design, energy sources, performance, and environmental impact. While conventional vehicles rely exclusively on internal combustion engines

(ICEs) powered by fossil fuels, HEVs integrate electric motors, battery packs, and sophisticated power electronics alongside the ICE to enhance efficiency, reduce emissions, and improve overall driving performance. The following discussion provides a detailed comparison of these two vehicle types, covering key operational parameters relevant to automotive engineering and energy management.HEVs are designed to use both an ICE and an electric motor to power the vehicle. The electric motor is powered by a battery that can be charged through regenerative braking and, in some cases, by plugging into an external power source. This combination allows the ICE to operate more efficiently, often shutting off during idling or low-speed driving and engaging only when higher power output is necessary. CVs, on the other hand, use a traditional ICE connected directly to the transmission system to deliver power to the wheels. While simpler in construction, this setup results in higher fuel consumption and emissions, especially under stop-and-go urban driving conditions.

Table 2.1: Comparison of Hybrid Electric Vehicles and Conventional Vehicles

Parameter	**Hybrid Electric Vehicles (HEVs)**	**Conventional Vehicles (CVs)**
Power Source	Combination of ICE and electric motor	Solely Internal Combustion Engine
Fuel Efficiency	Higher due to electric motor assistance	Lower due to complete reliance on ICE
Emissions	Significantly lower (reduced CO_2 & NOx)	Higher due to continuous fuel combustion
Initial Cost	Higher due to dual powertrain & battery system	Lower due to simpler construction
Maintenance Cost	Moderate; requires expertise for both systems	Relatively low due to established technology

Driving Range	Often higher due to regenerative braking and dual sources	Dependent on fuel tank capacity only
Technology	Advanced with regenerative braking & start-stop features	Traditional, mechanical-focused
Environmental Impact	More eco-friendly and sustainable	Higher carbon footprint

From the comparison table, it is evident that HEVs provide several advantages over CVs, particularly in terms of fuel efficiency, emission reduction, and environmental sustainability. HEVs are capable of recovering energy through regenerative braking, a feature absent in conventional vehicles. Moreover, their dual power source allows them to operate in electric-only mode during low-speed conditions, significantly reducing fuel consumption. However, these benefits come at the cost of higher initial investment, advanced maintenance requirements, and more complex engineering design.

Conversely, conventional vehicles offer simpler construction, lower manufacturing costs, and easier maintenance due to their mature and widely adopted technology. They remain dominant in regions where fuel costs are relatively low, and charging infrastructure for hybrid or electric vehicles is limited. However, growing concerns about greenhouse gas emissions and stringent environmental regulations are pushing the automotive industry toward hybrid and electric technologies, with HEVs serving as an important transitional technology between CVs and fully electric vehicles (EVs).

2.2 Impact of Modern Drive-Trains on Energy Supplies

The transition to modern electric and hybrid drivetrains has significantly influenced global energy supply dynamics. One of the primary advantages is the reduction in fossil fuel consumption, as vehicles equipped with electric assist features rely less on gasoline and diesel for propulsion. Additionally, plug-in hybrid electric vehicles (PHEVs) are capable of drawing power from the electricity grid, particularly during off-peak hours. This facilitates better load distribution and optimizes energy usage across the grid. Modern drive-trains, particularly in Hybrid Electric Vehicles (HEVs) and Plug-in Hybrid Electric Vehicles (PHEVs), are reshaping how energy is consumed and supplied in the transportation sector. One of the most significant impacts is the reduction in fossil fuel usage. Since electric motors assist the internal combustion engine, these vehicles consume less gasoline and diesel, which directly contributes to lower greenhouse gas emissions and decreased dependency on petroleum. Another important feature is the integration with the electrical grid. PHEVs can be charged using off-peak grid electricity, allowing for a more balanced energy demand curve and reducing the load during peak hours. This optimizes power plant efficiency and helps in integrating renewable energy sources like solar and wind. HEVs and PHEVs also utilize regenerative braking technology, which captures kinetic energy during braking and converts it into electrical energy to recharge the battery. This energy recovery system reduces the vehicle's total power demand, improving overall efficiency. However, this shift from fossil fuels to electrification introduces a new challenge: the increased demand for specific battery materials, namely lithium, nickel, and cobalt. These resources are critical for manufacturing high-performance batteries and are not as widely distributed globally as oil. This creates a shift in resource extraction focus and raises concerns about supply chain vulnerabilities and environmental impacts from mining.

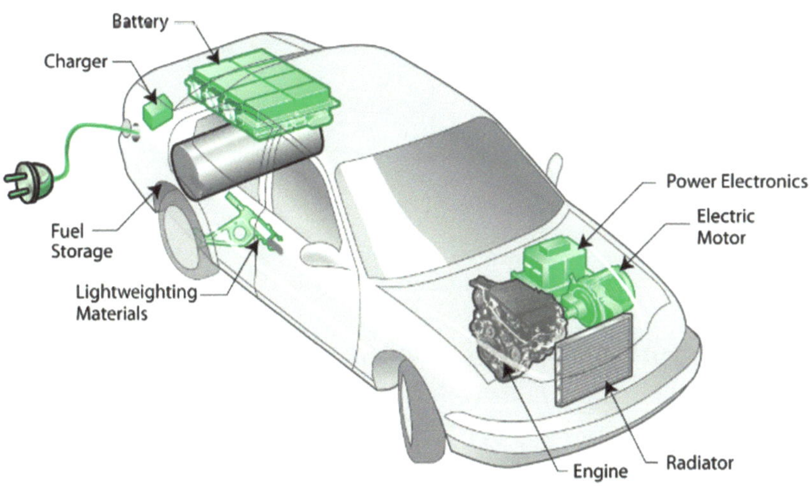

Fig 2.1: Component of PHEV

A Plug-in Hybrid Electric Vehicle (PHEV) is basically a hybrid electric vehicle (HEV) with an external charging option. Unlike conventional HEVs (which only charge the battery through

regenerative braking internal combustion engine), a PHEV can also be plugged into the electric grid to recharge its battery.

Plug-in HEV GTUexam-3170923-SUM2025

A Plug-in Hybrid Electric Vehicle represents a hybrid propulsion system in which an internal combustion engine operates in conjunction with an electric drive unit powered by a rechargeable energy storage system. This configuration allows the vehicle to utilize both conventional fuel and externally charged electrical energy for propulsion. Unlike traditional hybrids, PHEVs have the added advantage of being charged directly from an external power supply, such as a domestic outlet or a public charging station.

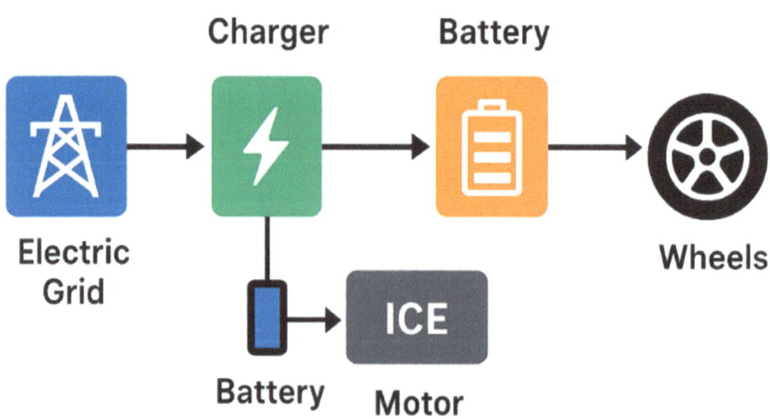

Fig 2.2: PHEV

This feature enables them to operate in pure electric mode for a longer distance, making them more eco-friendly and economical for short trips. The working principle of a PHEV involves three modes: in electric mode, the motor alone drives the vehicle using battery power; in hybrid mode, both the ICE and motor supply power together for improved efficiency and performance; and in charge-sustaining mode, the ICE maintains the battery's state of charge when it drops too low. PHEVs also use regenerative braking to recover energy during deceleration. Overall, they provide the flexibility of an EV for city driving and the extended range of an ICE for long-distance travel. So, a PHEV is basically a bridge between HEVs and fully electric vehicles (BEVs), giving the flexibility of fuel + electricity with extended range.

2.3 Hybrid Electric Drive-Trains

Hybrid electric drive-trains represent a significant advancement in the evolution of vehicle propulsion systems. These systems combine an internal combustion engine (ICE) with an electric motor and an onboard energy storage system, typically a battery. The resulting configuration allows the vehicle to operate using both fuel and electricity, optimizing energy use and reducing emissions depending on driving conditions. Drive train Configuration. An automotive drive train is shown in the Figure. It consists of a power plant, a clutch in a manual transmission or a torque converter in an automatic transmission, a gearbox, a final drive, a differential shaft, and driven wheels. A typical drive train layout, as illustrated in Figure, consists of several interlinked mechanical components, each playing a specific role in the propulsion system. At the core of the system is the power plant, usually an internal combustion engine (ICE) or an electric motor, which generates the mechanical energy required to move the vehicle. This power is then transmitted through a clutch in the case of manual transmissions, or a torque converter in automatic transmissions. These components act as couplers, managing the connection between the engine and the subsequent gear system while allowing for smooth engagement or disengagement of power. Following the clutch or torque converter, the gearbox (also known as the transmission) plays a key role in adjusting the torque and speed according to the driving conditions. It enables the engine to operate within an optimal speed range by selecting different gear ratios. Next in the sequence is the final drive, which further reduces the rotational speed and increases torque before delivering it to the differential shaft. The differential allows for the distribution of torque to the driven wheels, compensating for the speed difference between the inner and outer wheels during cornering, ensuring stability and smooth handling. Together, these components form an integrated drive train system that transforms engine output into controlled motion at the wheels, ensuring efficient torque transmission, vehicle responsiveness, and driver comfort.

Basic Concept of Hybrid Traction

The basic concept of hybrid traction is to revolves around the integration of an internal combustion engine (ICE), an electric motor, and a battery energy storage system within a single vehicle architecture. These hybrid powertrains are designed to utilize both conventional fuel (such as petrol or diesel) and electric energy to drive the vehicle, offering enhanced fuel efficiency, reduced emissions, and improved performance under varying operating conditions.

Depending on how the components are arranged and interact, hybrid vehicles can operate in one of the following primary modes: GTUexam-3170923-SUM2025

Series Hybrid Configuration: In this setup, the internal combustion engine is not mechanically connected to the wheels. Instead, it powers a generator, which either charges the battery or directly supplies electricity to the electric motor. The electric motor alone drives the wheels, allowing for smooth and efficient urban driving.

Parallel Hybrid Configuration: Here, both the ICE and the electric motor are connected to the wheels and can work either independently or in combination. This configuration allows the engine to contribute directly to propulsion, especially useful during highway driving when mechanical efficiency is highest. Series-Parallel (Power Split) Hybrid Configuration: This mode combines the advantages of both series and parallel systems. It features a power-splitting device that allows dynamic switching between series and parallel operation based on driving conditions. This flexible control strategy optimizes energy use, balancing performance with fuel economy and emissions reduction. This combination of components and operating strategies defines the core of modern hybrid traction systems, providing a balance between energy efficiency, performance, and sustainability.

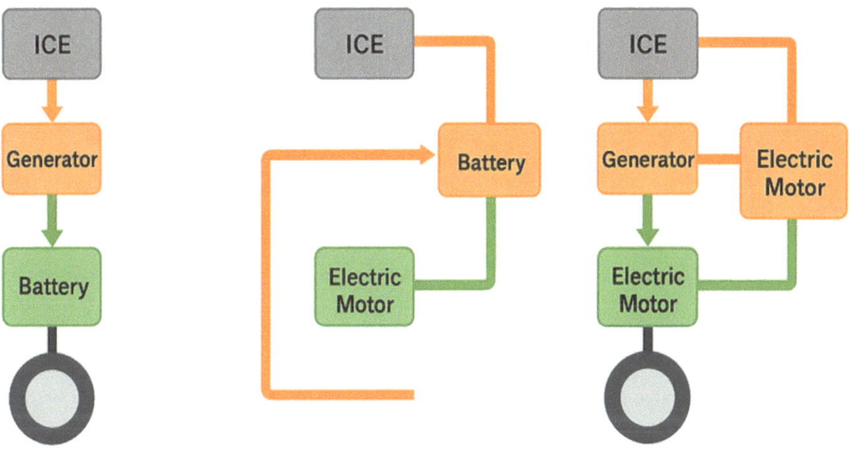

Fig 2.3: Series, parallel, series-parallel combination of Hybrid traction

2.4 Hybrid Drive-train Topologies

Hybrid electric vehicles (HEVs) are categorized based on how the engine, generator, motor, and battery are configured to deliver power to the wheels. These different drive-train topologies affect energy flow, fuel economy, control strategy, and overall performance. The most commonly implemented hybrid topologies are Series, Parallel, and Series-Parallel (Power Split) configurations.

In a Series Hybrid Topology, the internal combustion engine (ICE) is used exclusively to power a generator, which either charges the battery or supplies electricity to the electric motor. The electric motor is solely responsible for driving the wheels, meaning there is no direct mechanical connection between the engine and the wheels. This layout is particularly efficient in urban driving conditions and enables optimal ICE operation. An example of this architecture is the Chevrolet Volt.

2.4.1 The Parallel Hybrid Topology GTUexam-3170923-SUM2025

The Parallel Hybrid Topology enables both the engine and electric motor to drive the wheels directly. Power can be delivered by either system independently or in combination, making it highly efficient during steady-state cruising, where the ICE can operate in its most efficient range. This design also typically requires fewer components, making it lighter and simpler. The Honda Insight is a notable example of a parallel hybrid. The Series-Parallel (Power Split) Topology is a hybrid of the series and parallel configurations, allowing flexible control strategies. It uses a power-splitting device (such as a planetary gearset) to distribute power between the engine and the motor, optimizing energy use across various driving scenarios. This topology combines the low-speed efficiency of series hybrids with the high-speed performance of parallel hybrids. Widely adopted in the industry, this topology is used in vehicles like the Toyota Prius and Hyundai Ioniq.

2.4.2 Hybrid Electric Vehicle Architecture refers to the different topologies.

The integration of electromechanical components, including electric motors and generators, into automotive powertrains has renewed interest in the optimization of powertrain architectures. Growing regulatory and market pressure to improve fuel efficiency has intensified the need to fully exploit the potential of powertrain electrification. However, architectures currently available are typically tailored to specific vehicle categories, and their performance may not directly translate to vehicles with different mass, duty cycles, or operating conditions. For instance, a configuration optimized for passenger cars may fail to deliver acceptable fuel economy when applied to heavy-duty applications. Previous studies have demonstrated that powertrain architecture selection plays a critical role in determining overall fuel efficiency (Liu, 2007). Consequently, successful electrification strategies must rely on a systematic framework for selecting an architecture that aligns with vehicle characteristics and intended usage.

In hybrid electric vehicles, the powertrain architecture generally comprises an internal combustion engine, one or more motor–generator units, energy storage systems operating at both high and low voltage levels, power electronic components such as inverters and DC–DC converters, and a transmission that couples the propulsion sources to the vehicle drivetrain. From a system perspective, automotive powertrains can be categorized into electrical and mechanical architectures. The electrical architecture defines how batteries, power electronics, and motor–generators are interconnected along the electrical energy path, whereas the mechanical architecture describes the physical arrangement and coupling of the engine, motor–generators, transmission, and final drive. This dissertation focuses specifically on the mechanical architecture.

A mechanical powertrain architecture may incorporate multiple configurations that allow transitions between different operating states during vehicle operation. Within a hybrid electric vehicle, each of these configurations is commonly referred to as a driving mode. Accordingly, an HEV architecture can be interpreted as a structured set of available driving modes designed to enhance efficiency and drivability under varying operating conditions. One widely adopted classification approach distinguishes these modes based on the manner in which the engine and electric machines are interconnected. Using this criterion, HEV configurations are typically categorized into three fundamental classes: series, parallel, and power-split architectures.

Series Configuration

The series configuration owes its name to the components being connected in a series manner, see Figure An engine is connected to a generator that charges the battery. The battery is also connected to multiple motors that drive the vehicle. This design provides simplicity in the architecture. In this configuration, since the only purpose of the engine is to charge the battery, the engine runs independently of the vehicle's motive operation. That way the engine can run in its efficient operating region regardless of the vehicle's motive operation. However, energy conversion at the generator and motor stages is a major source of energy loss. Also, at high speeds electric motors do not run as efficiently as internal combustion engines, which makes such a design not suitable for highway applications but a good alternative for city driving. A series hybrid architecture, i.e., an architecture consisting of only a series hybrid mode has not been implemented in any of the commercially produced hybrid vehicles but the Chevrolet Volt has a series hybrid mode in addition to other driving modes.

Parallel Configuration GTUexam-3170923-SUM2025

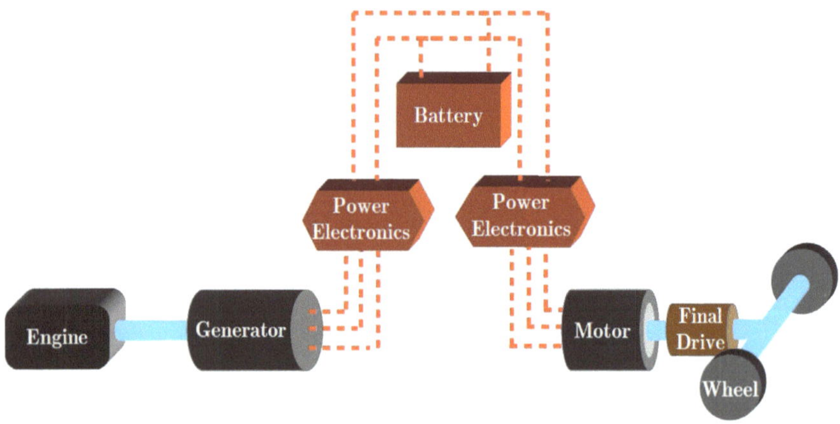

Fig 2.4: Parallel Configuration

In the parallel configuration, an engine and an MG drive the vehicle together, properly connected by a transmission. See Figure. The speeds of the MG and engine is linked via gearing and torques from both sources are combined to drive the vehicle. The MG is also used to charge the battery in which case the engine drives both the vehicle and motor together. In order to obtain a variety of operating regions a parallel hybrid configuration typically includes either a conventional automatic transmission or a continuously variable transmission. A parallel hybrid drive train architecture features both an Internal Combustion Engine (ICE) and an electric motor (EM) mechanically coupled to the same transmission, allowing them to independently or simultaneously provide power to the vehicle's wheels. Power flow is controlled by a vehicle controller, which manages the system's four main operating modes:

All-Electric Mode (EM powers the vehicle), Engine-Only Mode (ICE powers the vehicle), Hybrid Mode (both ICE and EM share power), and Regenerative Braking Mode (EM acts as a generator to charge the battery). The Integrated Motor-Assist System, developed by Honda Motor Company, uses this architecture in mass-produced vehicles such as the Honda Insight Hybrid (Aoki et al., 2000).

Power-split Configuration

The power-split powertrain relies fundamentally on the use of a planetary gear set, which enables flexible power flow management while maintaining high transmission efficiency. Within this arrangement, the internal combustion engine and motor–generator units are mechanically coupled through one or more planetary gear sets to collectively propel the vehicle, as illustrated in Figure. A typical planetary gear mechanism consists of a centrally located sun gear, an outer ring gear, and a set of intermediate planet gears that mesh with both.

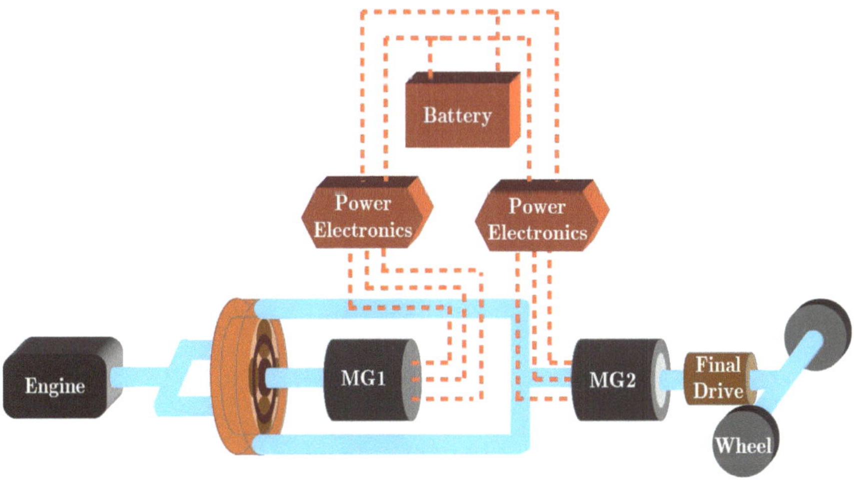

Fig 2.5: Power-Split Hybrid Architecture

The pinions are connected via a bar, altogether forming the carrier gear as shown on Figure.

Depending on the number of PGs in the architecture, there are many possible ways to connect the engine, motors, and vehicle output, providing many alternative configurations to select from. Power-split configurations that have been proposed in the literature include 1-PG, 2-PG, and even 3-PG systems, see Table. Commercial implementations of this configuration include the Toyota Prius and the Ford C-Max.

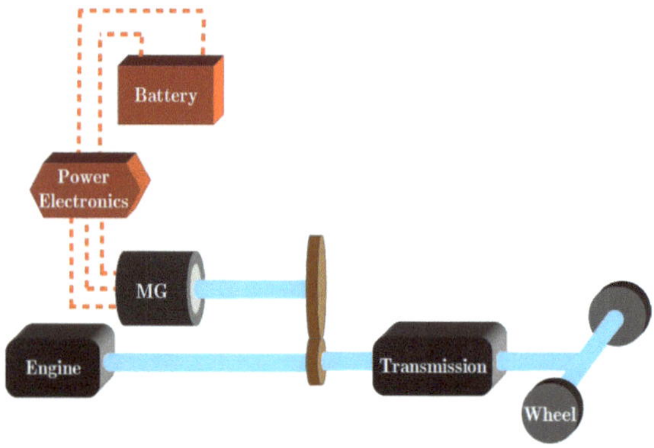

Fig 2.6: Parallel Hybrid Architecture

Also, one of the modes of the Chevrolet Volt is a power-split configuration (Conlonet al., 2011).

Table 2: Existing architectures from the literature

Number of PGs	Reference
1-PG	*Conlon et al.* (2011) *Sasaki* (1998)
2-PG	*Schmidt* (1996b) *Schmidt* (1996a) *Holmes et al.* (2003) *Holmes and Schmidt* (2002) *Ai and Anderson* (2005)
3-PG	*Schmidt* (1999b) *Raghavan et al.* (2007)

Planetary gear (PG) sets enable the generation of a wide range of powertrain architecture alternatives. For instance, parallel configurations can be realized by grounding specific gears within a PG set. Owing to this inherent flexibility, this dissertation primarily concentrates on architectures formed through different configurations of powertrain components using PGs. In the absence of a parametric representation capable of defining the complete architecture design space, it becomes essential to employ a modelling approach that can systematically represent all possible architecture alternatives. To address this requirement, a novel modified bond graph representation has been developed. Chapter 3 presents this representation in detail and introduces a structured framework for generating all feasible configuration alternatives.

Comparison of series hybrid and parallel hybrid systems with their merits and Demerits GTUexam-3170923-WIN2024

Table 2.2: Comparison of series hybrid and parallel hybrid systems

Feature	Series Hybrid	Parallel Hybrid
Power Flow	Engine → Generator → Motor → Wheels	Engine + Motor both directly drive wheels
Complexity	Simpler mechanical, more electrical	More complex mechanical couplings
Efficiency	Low at high speed (double conversion losses)	Higher at highway speeds
Performance	Better low-speed performance (motor torque)	Good balance at all speeds
Battery Dependence	High (motor always needs battery support)	Moderate (engine can drive wheels directly)
Cost	Higher (due to larger motor & generator)	Moderate
Applications	Buses, range-extender EVs	Passenger cars, mild hybrids

2.5 Power Flow Control in Hybrid Drive-Trains

Hybrid electric vehicles (HEVs) are equipped with advanced power management systems that dynamically control the flow of energy between the internal combustion engine (ICE), electric motor, generator, and battery based on driving conditions. One of the most fundamental aspects of hybrid vehicle functionality is the ability to switch between various power flow modes to optimize efficiency, performance, and emissions. In EV mode, the vehicle is powered solely by the electric motor using energy from the battery, with the ICE turned off. This mode is ideal for low-speed urban driving, where noise and emissions need to be minimized. In series mode, the ICE does not directly drive the wheels; instead, it powers a generator that charges the battery or supplies electricity to the motor. This setup is particularly useful when the battery's state of charge (SOC) is low. Parallel mode enables the ICE and the electric motor to work together, simultaneously delivering torque to the wheels. This configuration is often used during acceleration or climbing, where higher power output is required. During regenerative braking, the electric motor operates in reverse as a generator, converting kinetic energy into electrical energy, which is then stored in the battery. This significantly improves the vehicle's overall energy efficiency by reclaiming energy that would otherwise be lost as heat. Lastly, in engine-only mode, typically engaged at steady high speeds, the ICE alone powers the vehicle to maintain efficiency and reduce energy conversion losses. These multiple modes of power flow not only allow for flexible operation under varying driving conditions but also contribute to reduced fuel consumption and lower greenhouse gas emissions.

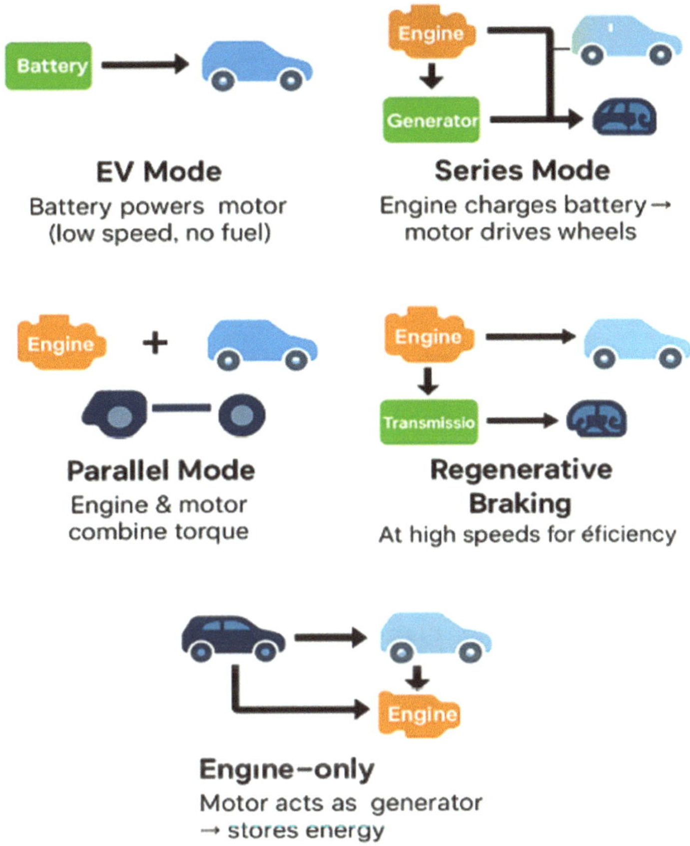

Fig 2.7: Power flow control in Hybrid drive train

2.6 Power Electronics Control Units in Hybrid Vehicles

In hybrid and electric vehicles, Power Electronics Control Units (PECUs) play a critical role in determining and executing optimal power flow strategies. These units act as intelligent decision-makers, continually evaluating several key parameters to switch between driving modes such as EV, series, parallel, or engine-only. The most influential parameter is the State of Charge (SOC) of the battery; a low SOC typically triggers the engine to assist or charge the battery. Vehicle speed and load demand further influence whether the system engages the motor alone or combines engine power for higher torque requirements. Inputs from the accelerator and brake pedals are also monitored in real-time to assess driver intent, such as sudden acceleration or deceleration, which affects energy recovery or power boost strategies. Additionally, modern hybrids must meet stringent emission targets, requiring the system to reduce ICE usage during low-efficiency zones or switch to electric-only operation in emission-restricted areas. The decision-making process is orchestrated by multiple subsystems, including the Electronic Control Units (ECUs), Battery Management System (BMS), and Inverter Controllers, which communicate via a high-speed network to ensure seamless power delivery, safety, and efficiency.

2.6.1 Economic and environmental impact of electric hybrid vehicle GTUexam-3170923-SUM2025WIN2024

Electric hybrid vehicles (HEVs) offer significant economic and environmental benefits. Economically, they reduce fuel consumption, lower operating and maintenance costs, and decrease dependence on fossil fuels, making them cost-effective over their lifespan. Environmentally, HEVs help reduce greenhouse gas emissions and air pollutants by combining electric propulsion with efficient internal combustion engines. Regenerative braking and optimized energy management further enhance efficiency, contributing to cleaner air, reduced carbon footprint, and a more sustainable transportation system.

2.6.2 Fuel Efficiency Analysis of Hybrid Vehicles

Table 2.3: Efficiency analysis of ICE & HEV

Drive Condition	ICE	HEV
City Traffic	Low efficiency	High efficiency
Highway Drive	Moderate	Slightly better
Start-Stop Traffic	Very poor	Excellent (regen braking, EV mode)
Hill Climb	High fuel use	Motor assist reduces load

Efficiency Calculation

Fuel economy represents how efficiently a vehicle uses fuel to cover a certain distance. It is usually expressed in kilometres per liter (km/l).

$$\text{Fuel Economy} = \frac{\text{Distance Traveled}}{\text{Fuel Consumed}} \quad (\text{e.g, km/l})$$

Distance Traveled   Fuel Consumed

This value is important when comparing the performance of different vehicles under similar driving conditions.

Effective Fuel Use

Hybrid vehicles like the Toyota Prius also utilize regenerative braking, which recovers kinetic energy during braking and converts it into electrical energy stored in the battery. This recovered energy reduces the effective fuel consumption.

$$\text{Effective Fuel Use} = \text{Fuel Energy} + \text{Regenerated Energy from Brakes}$$

Thus, the total usable energy comes from both: Chemical energy of the fuel and regenerative braking energy

Example: Toyota Prius Gen 4

The fourth-generation Toyota Prius is known for its high efficiency due to its advanced hybrid powertrain and regenerative braking system.

Real-world City Mileage:

$$\approx 28 \text{ to } 30\,\text{km}/\text{l}$$

This efficiency reflects both traditional fuel use and the contribution from regenerative braking, making it more economical than conventional ICE vehicles in urban driving.

Remembrance: Social & Environment Impact- Lower CO_2 and better urban air quality, Energy Impact- Shifts demand from oil to grid/batteries, Hybrid

Traction- Combines ICE and electric drive, Topologies- Series, Parallel, Series-Parallel, Power Flow- Controlled via ECUs, BMS, Efficiency- Best in stop-go city traffic, regen braking saves energy.

2.6.3 Anti-lock Braking System (ABS)-ABS means an electronically controlled regenerative braking system GTUexam-3170923-SUM2025

An electronically controlled regenerative braking system functions in coordination with an Anti-lock Braking System (ABS), not as an ABS itself, to ensure vehicle stability and energy efficiency during braking. The system integrates both regenerative and hydraulic friction brakes, using a sophisticated controller to manage the vehicle's braking torque. During heavy or emergency braking, the controller monitors wheel speed to prevent lock-up, adjusting the distribution between regenerative and friction braking to maintain optimal wheel slip and steering control, similar to a conventional ABS, while also recovering kinetic energy for the battery.

How it Works

1. **ABS Functionality First:**

   At the core of the system is the ABS, which constantly monitors wheel speeds to detect impending wheel lock-up.

2. **Brake Torque Allocation:**

   A sophisticated controller decides how much braking force to apply. It prioritizes using electric motor acting as a generator for regenerative braking to capture energy.

3. **Coordinated Action:**

   - **Normal Braking:** The system primarily uses regenerative braking to recover energy, slowing the vehicle.
   - **Heavy/Emergency Braking:** If the ABS detects that the wheels are about to lock up, it will modulate the amount of regenerative

braking applied. It can also blend in hydraulic (friction) brakes as needed to achieve the desired stopping power while keeping the wheels rotating, preventing skidding.

4. **Energy Recovery:**

   When regenerative braking is used, the electric motor converts the vehicle's kinetic energy into electrical energy, which is then stored in the vehicle's battery.

5. **Stability and Control:**

   By ensuring the wheels don't lock up, the system maintains directional stability and the driver's ability to steer during hard braking, even on slippery surfaces.

Key Aspects

- **Blended Systems:**

  Rather than relying solely on regenerative braking, the system blends it with conventional friction brakes.

- **Smart Controllers:**

  Advanced control algorithms, such as fuzzy logic or model predictive control, are used to manage the complex coordination between the regenerative and hydraulic braking systems.

- **Energy Efficiency:**

  The primary goal is to recover as much kinetic energy as possible for storage in the battery, which increases the vehicle's driving range.

- **Safety:**

  By functioning like a conventional ABS, the system prevents accidents by ensuring the vehicle remains stable and controllable under all braking conditions.

2.7 Communication & Networking in HEVs and EV for Vehicle Control GTUexam-3170923-SUM2025

The Controller Area Network (CAN) is a serial communication protocol developed for real-time communication in automobiles. It allows different electronic control units (ECUs) in HEVs/EVs to exchange information reliably without a central computer.

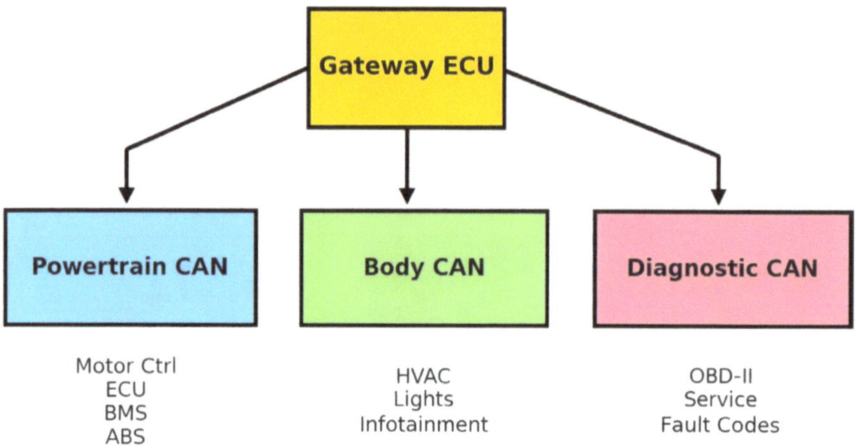

Fig 2.8: Controller Area Network (CAN)

A typical CAN network in HEVs is divided into different functional domains for efficient communication. The Powertrain CAN connects critical components such as the motor controller, battery management system (BMS), engine control unit (ECU), transmission, and regenerative braking system. It handles high-priority data, including torque requests, motor speed, battery voltage, state of charge (SOC), and regenerative braking signals, ensuring smooth powertrain operation. The Powertrain CAN in HEVs connects major drivetrain components such as the motor controller, battery management system (BMS), engine control unit (ECU), transmission, and regenerative braking system. It is responsible for handling high-priority data, including torque requests, motor speed, battery voltage, state of charge (SOC), and regenerative braking signals.

By ensuring fast and reliable communication between these critical units, the Powertrain CAN enables efficient energy management, smooth acceleration, and safe braking in hybrid and electric vehicles. The Body CAN manages auxiliary systems such as lighting, HVAC, door controls, and infotainment, and it deals mainly with medium-priority data related to comfort and convenience. The Diagnostic CAN (OBD-II) is used primarily for service and maintenance, providing information on vehicle fault codes, charging status, and performance data to external diagnostic tools. To integrate all these networks, the Gateway ECU serves as a communication bridge between Powertrain CAN, Body CAN, and Diagnostic CAN, enabling secure and prioritized data transfer across the entire hybrid vehicle system.

Multiple Choice Questions

1. **Which of the following is an advantage of hybrid electric vehicles over conventional ICE vehicles?**

   A) Higher emissions

   B) Lower fuel economy

   C) Regenerative braking

   D) Complete reliance on fossil fuel

   Correct Answer: C) Regenerative braking Explanation:HEVs recover energy during braking, improving efficiency.

2. **In a series hybrid vehicle, the internal combustion engine (ICE):**

   A) Drives the wheels directly

   B) Charges the battery or powers the motor via generator

   C) Is not used at all

   D) Acts only during braking

   Correct Answer: B) Charges the battery or powers the motor via generator

   Explanation: In series configuration, ICE is decoupled from the wheels.

3. **Which component manages power flow between battery, motor, and engine in an HEV?**

   A) ECU

   B) Fuel Tank

   C) Differential

   D) Radiator

   Correct Answer: A) ECU Explanation: The Electronic Control Unit (ECU) governs energy distribution in hybrids.

4. **What is the main environmental benefit of using hybrid vehicles?**

    A) Increased vehicle noise
    B) Higher CO_2 emissions
    C) Reduced greenhouse gas emissions
    D) Depletion of battery resources

    Correct Answer: C) Reduced greenhouse gas emissions

    Explanation: HEVs emit less CO_2 than pure ICE vehicles.

5. **Which of the following is a Series-Parallel hybrid vehicle?**

    A) Honda Civic
    B) Chevrolet Volt
    C) Toyota Prius
    D) Maruti 800

    Correct Answer: C) Toyota Prius

    Explanation: Prius uses a blend of series and parallel drive for better efficiency.

6. **What is the major limitation of hybrid vehicles compared to BEVs?**

    A) Use of regenerative braking
    B) Partial dependence on fossil fuel
    C) Lower fuel economy
    D) Fully electric operation only

    Correct Answer: B) Partial dependence on fossil fuel

    Explanation: Unlike BEVs, hybrids still use an ICE.

7. **Which drive-train configuration offers maximum flexibility in power management?****

    A) Series
    B) Parallel
    C) Series-Parallel
    D) All have same efficiency

    Correct Answer: C) Series-Parallel

    Explanation: Combines benefits of both series and parallel topologies.

8. What is the role of regenerative braking in HEVs?

A) Wastes kinetic energy

B) Converts heat to electricity

C) Stores mechanical energy

D) Converts kinetic energy to electrical energy

Correct Answer: D) Converts kinetic energy to electrical energy

Explanation: Energy is stored in the battery during braking.

9. Which factor does NOT impact HEV fuel efficiency significantly?

A) Driving cycle

B) Road gradient

C) Paint color

D) Traffic density

Correct Answer: C) Paint color

Explanation: Aerodynamics matter, but colour has a negligible impact.

10. In HEVs, the electric motor:

A) Can never run alone

B) Only drives the wheels in parallel mode

C) Can assist or fully drive the wheels depending on mode

D) Is just used to start the engine

Correct Answer: C) Can assist or fully drive the wheels depending on mode

Explanation: In EV or hybrid mode, electric motor can operate independently.

India-GTU Questions

List the economic and environmental impact of electric hybrid vehicles.	chapt2pg52SUM2025	
What is a plug-in HEV? Explain with a figure.	chapt2pg40	
Draw and Explain typical CAN network for HEVs.	chapt2pg56	**07**
Draw and explain the architecture and power flow control of a parallel hybrid electric drive train.	Chapt2pg36WIN2024	**07**
Give comparisons of Hybrid Electric Vehicles and Conventional Vehicles.	chapt2 pg36,37	**03**
Compare series hybrid and parallel hybrid systems with their merits and demerits. 07	Chapt2pg42,49	
Explain Electronically controlled regenerative braking system functioning as an ABS.	chapt2 pg25 pg29SUM2025	**07**
What is the need and importance of electric and hybrid electric vehicles?	Chapt2 pg 28WIN2025	**03**
Compare hybrid electric vehicles with conventional vehicles.	Chapt2pg36,37 WIN2025	**03**
Compare series hybrid and parallel hybrid systems with their merits and demerits.	Chapt2pg42,49	**07**

Chapter 03

Electric Drive Trains and Propulsion Units

3.1 Basic Concept of Electric Traction GTUexam-3170923-SUM2025

Electric drive train subsystems, focusing on the drive train. It explains the common parts between electric vehicles (EVs) and petrol vehicles, such as the body, frame, doors, power windows, wheels, and suspension system. The lesson also highlights the parts unique to EVs, including the electric motor, controller, battery pack, and vehicle control unit. It further discusses the impact of parameters like rolling resistance, aerodynamic coefficient, vehicle frontal area, weight, slope, pickup acceleration, and regeneration on the design of EVs.

Electric traction refers to the use of electric motors for vehicle propulsion. It is a core technology in modern transportation systems such as electric cars, electric trains, metros, and trams. The principle of electric traction involves converting electrical energy into mechanical energy to drive the vehicle's wheels. This process is achieved through the coordinated functioning of four primary components: the traction motor, the controller, the energy source, and the transmission system. The traction motor is the central component of the traction system. It is responsible for converting electrical energy supplied by a battery or external source into mechanical torque. This torque is then used to rotate the vehicle's wheels. Various types of motors are employed in electric traction, including DC motors, induction motors, brushless DC motors (BLDC), and permanent magnet synchronous motors (PMSM). These motors are preferred due to their high starting torque, energy efficiency, and precise controllability, especially in stop-and-go urban transport. The controller plays a critical role in managing the power flow from the energy source to the motor. It regulates motor functions such as speed, torque, and direction of rotation, based on inputs from the driver (such as accelerator and brake commands) and system feedback (such as vehicle load, gradient, and battery state of charge). Advanced controllers use technologies like pulse-width modulation (PWM) and inverter circuits for AC motor control. Moreover, the controller is responsible for implementing regenerative braking, where the motor acts as a generator to convert kinetic energy back into electrical energy during deceleration. The energy source supplies the electrical energy required by the system. In electric vehicles, this source is commonly a lithium-ion battery, which provides direct current (DC). In contrast, electric rail systems often draw alternating current (AC) from overhead lines or third rails. Alternative sources such as hydrogen fuel cells are also being adopted in specific applications. The energy source must offer sufficient energy density, fast charging capability, and thermal stability to meet the demands of the drive system. Finally, the transmission system is responsible for transferring mechanical torque from the motor to the vehicle's wheels. Unlike internal combustion engine (ICE) vehicles that require multi-speed gearboxes, electric vehicles often use a single-speed reduction gear or direct drive due to the broad torque-speed characteristics of electric motors. The transmission includes components such as the final drive, differential, and axles, ensuring smooth and efficient power delivery to the road. Together, these components form a seamless and efficient propulsion system. Electric traction systems are increasingly being adopted worldwide due to their lower environmental impact, higher efficiency, and suitability for automation and intelligent control systems.

3.1.1 Components of Hybrid Electric Vehicle (HEV) Drive Train GTUexam-3170923-SUM2025

Components and Advantages of Hybrid Electric Vehicles (HEVs)

Hybrid Electric Vehicles (HEVs) integrate both internal combustion engines (ICE) and electric powertrain components to improve fuel efficiency, reduce emissions, and enhance driving performance. The drive train of an HEV includes various electrical and mechanical systems that work together to transmit power from the energy sources to the wheels. This section explains the major components of an HEV drive train and highlights the advantages of hybrid vehicle technology. A Hybrid Electric Vehicle (HEV) drive train integrates conventional internal combustion engine (ICE) components with electric propulsion systems to achieve improved fuel efficiency and reduced emissions. The primary components are:

Internal Combustion Engine (ICE): Provides mechanical power using fuel and operates primarily during high load or extended driving conditions. Electric Motor/Generator: Functions as a motor during acceleration and as a generator during regenerative braking. Battery Pack: Stores electrical energy to power the electric motor, charged via regenerative braking or ICE. Power Electronics Controller: Manages power flow between the battery, electric motor, and ICE to ensure optimal performance. Transmission System: Transfers power from the ICE and/or electric motor to the wheels; may include CVT or e-CVT. Regenerative Braking System: Captures kinetic energy during braking and converts it to electrical energy for storage. Fuel Tank and Fuel Delivery System: Stores and delivers fuel to the ICE when required. The Internal Combustion Engine (ICE) in a Hybrid Electric Vehicle (HEV) serves as the primary or supplementary power source, depending on the drive mode. It generates mechanical energy by burning conventional fuels such as petrol or diesel. In HEVs, the ICE is often optimized to operate within its most efficient range to minimize fuel consumption and reduce emissions. When not required, the engine can be shut down, allowing the electric motor to take over, thereby improving overall vehicle efficiency.

An electric motor/Generator functions both as a motor (driving the vehicle) and as a generator (regenerating energy during braking). Their types are Permanent Magnet Synchronous Motor (PMSM), Induction Motor (IM). The Electric Motor/Generator in a Hybrid Electric Vehicle (HEV) performs a dual function. As a motor, it converts electrical energy from the battery into mechanical energy to drive the wheels, providing smooth and instant torque. As a generator, it converts mechanical energy (from the wheels or engine) back into electrical energy during regenerative braking or while coasting, thereby recharging the battery. Common types of electric motors used in HEVs include Permanent Magnet Synchronous Motors (PMSM) and Induction Motors (IM), selected for their high efficiency, compact size, and excellent torque characteristics.

The Energy Storage System (Battery Pack) stores electrical energy to power the electric motor. There are Common types: Lithium-ion (Li-ion), Nickel-Metal Hydride (NiMH).

Power Electronics Controller (PEC) manages power flow between the battery, motor, and generator, which includes DC-DC converters and inverters.

The Transmission System transfers power from the ICE and/or electric motor to the wheels. There are two types: Automated Manual Transmission (AMT) and Continuously Variable Transmission (CVT).

Control Unit (Hybrid Control Module – HCU) supervises energy management strategies and optimizes ICE and motor usage for efficiency and performance.

Regenerative Braking System converts kinetic energy during braking into electrical energy and stores recovered energy in the battery.

Fuel Tank stores conventional fuel (Petrol/Diesel) for ICE operation.

Auxiliary Components are cooling systems, onboard chargers, DC-DC converters for auxiliary loads, etc.

3.1.2 Advantages of Hybrid Vehicles GTUexam-3170923-WIN2024

- Better Fuel Efficiency – Electric motor assists in low-speed driving, reducing overall fuel consumption.
- Lower Emissions – Reduced greenhouse gas and pollutant emissions compared to conventional vehicles.
- Instant Torque and Smooth Acceleration – Electric motor provides immediate power response for better performance.
- Extended Driving Range – A Combination of fuel and electric systems allows longer travel distances.
- Energy Recovery via Regenerative Braking – Captures wasted kinetic energy and reuses it to charge the battery.
- Reduced Operating Costs – Lower fuel consumption and improved efficiency reduce overall running expenses.

3.1.3 Comparison of Lithium-Ion and Lead Acid Battery Configurations in EV GTUexam-3170923-SUM2025

Lithium-ion batteries offer higher energy density, lighter weight, faster charging, and longer lifespan than lead-acid batteries, making them the preferred choice for modern electric vehicles despite their higher upfront cost. Lead-acid batteries are cheaper and easier to install, but are heavier, have lower capacity, and significantly shorter lifespans, severely limiting their use in contemporary EVs.

Here is a detailed comparison:

Lithium-ion Batteries:

- **Performance:** Have higher energy density, providing more power and a longer range from a smaller and lighter pack.
- **Lifespan:** Features a much longer cycle life (5,000 cycles) and a longer overall operational lifespan compared to lead-acid batteries.

- **Charging:** Can accept faster charging rates and can be partially charged without affecting the battery's life.
- **Usable Capacity:** Can deliver almost their full rated capacity, even under high loads.
- **Environmental Impact:** Considered more environmentally friendly, with fewer hazardous materials and a smaller overall environmental footprint.
- Cost: Have a higher upfront cost.
- **Maintenance:** Requires specific charging procedures and may be sensitive to extreme temperatures.

Lead-Acid Batteries:

- **Performance:** Have lower energy density, making them heavier and larger for the same amount of power.
- **Lifespan:** Offers a much shorter cycle life (300-500 cycles) and a significantly shorter overall lifespan, with deep discharges negatively impacting their life.
- **Charging:** Charging takes longer, and deep cycling significantly reduces their lifespan.
- **Usable Capacity:** Offers substantially less usable energy under high loads, with capacity often reduced by 50% or more.
- **Environmental Impact:** Contains hazardous materials like sulfuric acid, posing higher environmental risks.
- **Cost:** Have a lower upfront cost, making them an option for budget-conscious or shorter-term applications.
- **Maintenance:** Require more maintenance and are sensitive to extreme temperatures and improper storage

Fig 3.1: Lead-acid Versus Lithium-ion Battery

As lead-acid and Lithium-ion batteries are separated into different groups as mentioned above, we will take flooded lead-acid, VRLA lead-acid and Lithium-ion (LiNCM) for comparison.

Cycle life

3.1.4 Comparative Performance of Lithium-Ion and Lead-Acid Batteries

Cycle Life and Discharge Characteristics

Table 3.1: Comparison of Lead-acid Versus Lithium-ion battery

Parameters	Flooded lead-acid	VRLA Lead-acid	Lithium-ion
Energy Density (Wh/L)	80	100	250
Specific Energy (Wh/L)	30	40	150
Regular Maintenance	Yes	No	No
Initial Cost ($/kWh)	65	120	600

Cycle life	1,200 @ 50%	1,000 @ 50% DoD	1,900 @ 80% DoD
State of Charge window	50%	50%	80%
Temperature sensitivity	degrades above 25 degrees	degrades above 25 degrees	degrades above 45 degrees
Voltage Increments	2V	2V	2V
Efficiency	60% @ 1 hr rate	60% @ 1 hr rate	92% @ 1 hr rate

Lithium-ion batteries exhibit a substantially higher cycle life than lead-acid batteries in discharge-based applications. While the cycle life of both chemistries can be enhanced by controlling depth of discharge (DoD), discharge rate, and operating temperature, lead-acid batteries are significantly more sensitive to variations in these parameters. As a result, lithium-ion systems maintain performance and longevity more effectively under comparable operating conditions.

Rate Performance

Lithium-ion batteries demonstrate superior rate performance when compared to lead-acid batteries. For instance, a 100 Ah valve-regulated lead-acid (VRLA) battery typically delivers only about 80 Ah when discharged over four hours. In contrast, a 100 Ah lithium-ion battery can deliver more than 98 Ah even during a high-rate, 30-minute discharge. This characteristic makes lithium-ion batteries particularly suitable for applications where full discharge occurs within eight hours, such as high-power appliances and fast-cycling energy systems.

Cold Weather Performance

Both lithium-ion and lead-acid batteries experience capacity reduction at low temperatures. However, lithium-ion batteries retain a higher proportion of their nominal capacity as temperatures decrease, maintaining relatively better performance even at temperatures as low as −20 °C. This provides lithium-ion systems with a clear advantage in cold-climate operations.

Environmental Impact

From an environmental perspective, lead-acid batteries perform less favorably compared to lithium-ion batteries. Lead-acid battery manufacturing requires a greater quantity of raw materials to achieve the same energy storage capacity, resulting in increased environmental degradation during mining and processing. Furthermore, the production and disposal processes associated with lead-acid batteries generate significant pollution, posing serious risks to ecosystems and human health.

Safety Considerations

In both battery types, failure conditions can result in rapid temperature rise, electrolyte leakage, fire, and the release of hazardous fumes. However, lithium-ion batteries present a comparatively higher safety risk due to their high energy density, which concentrates a large amount of stored energy within a smaller physical volume. Consequently, thermal runaway events in lithium-ion systems can be more severe if not properly managed.

3.2 Classification of Electric Motor Drives for EV and HEV Applications GTUexam-3170923-SUM2025

Electric motor drives are essential components in Electric Vehicles (EVs) and Hybrid Electric Vehicles (HEVs), as they determine the vehicle's performance, efficiency, and control. These drives convert electrical energy from batteries or fuel cells into mechanical energy to drive the wheels. The classification of motor drives is based on the type of motor, control strategy, and power converter topology.

Electric motor drives are essential components in Electric Vehicles (EVs) and Hybrid Electric Vehicles (HEVs), as they determine the vehicle's performance, efficiency, and control. These drives convert electrical energy from batteries or fuel cells into mechanical energy to drive the wheels. The classification of motor drives is based on the type of motor, control strategy, and power converter topology.

Electric motor drives play a vital role in Electric Vehicles (EVs) and Hybrid Electric Vehicles (HEVs) by converting electrical energy from batteries or fuel cells into mechanical energy to propel the vehicle. They are responsible for delivering the required torque, speed, and performance while ensuring high efficiency and reliability. The classification of electric motor drives is generally done based on the type of motor used, the control strategy adopted, and the power converter topology.

Based on the type of motor, electric drives are categorized into DC motor drives and AC motor drives. DC motor drives include Brushed DC Motors (BDC) and Permanent Magnet DC Motors (PMDC). BDC motors are simple in control and provide high starting torque but require periodic maintenance due to the presence of brushes. PMDC motors are more efficient and compact due to the use of permanent magnets, making them ideal for light-duty EV applications. AC motor drives comprise Induction Motors (IM), Permanent Magnet Synchronous Motors (PMSM), Brushless DC Motors (BLDC), and Switched Reluctance Motors (SRM). Induction motors are rugged, reliable, and widely used in commercial EVs such as the Tesla Model S. PMSMs are known for their high efficiency, compact design, and excellent torque density. BLDC motors combine the advantages of DC control with AC operation, offering high efficiency and low maintenance. SRMs, with their simple construction and high reliability, are gaining attention but require advanced control techniques to reduce torque ripple.

Based on the control strategy, motor drives can be classified as Scalar Control, Vector Control, or Direct Torque Control (DTC). Scalar Control, commonly referred to as V/f control, maintains a constant voltage-to-frequency ratio and offers simplicity but limited dynamic performance. Vector Control, also known as Field-Oriented Control (FOC), provides independent control of flux and torque, resulting in smooth torque production and higher efficiency. Direct Torque Control

(DTC) provides fast torque response and superior performance, making it suitable for high-end electric vehicles.

Finally, based on power converter topology, electric motor drives can be divided into DC–DC converters, DC–AC inverters, and bidirectional converters. DC–DC converters are used for voltage regulation, DC–AC inverters supply power to AC motors, and bidirectional converters enable regenerative braking and energy recovery during deceleration, thereby improving overall vehicle efficiency.

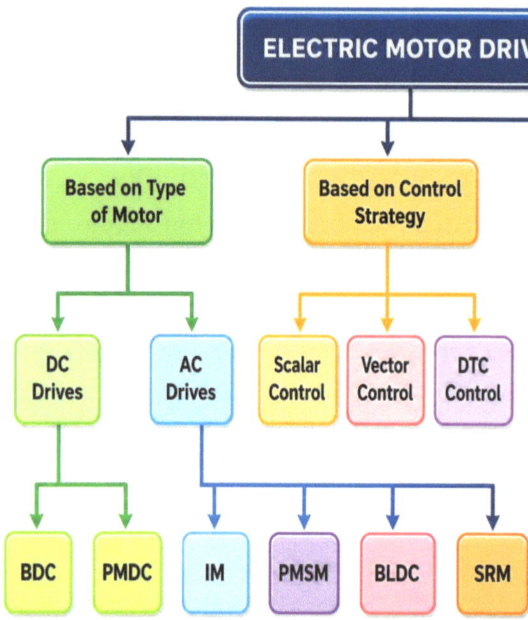

Fig 3.2: Classification of Motor Drives

3.3 Operation & Controlling of DC Motor Drive GTUexam-3170923-SUM2025

A DC motor drive converts electrical energy into mechanical torque by controlling the voltage and current applied to the motor. The power converter, such as a rectifier or chopper, regulates the armature voltage for speed control and the field current for torque and flux control. In motoring mode, energy flows from the supply to the motor, while in regenerative braking, energy flows back to the supply. By varying the armature voltage, speed is controlled below the base speed, and by weakening the field current, operation above base speed is achieved. Closed-loop control with sensors, current controllers, and speed controllers ensures stable performance and four-quadrant operation, allowing forward/reverse motoring and braking. A DC motor drive controls a DC motor's operation by adjusting the voltage and current supplied to its armature and field windings, enabling control of starting, stopping, reversing, speed, and torque. The drive, a functional circuitry, converts AC power to controllable DC, using methods like a rectifier for voltage control and a pulse width modulator or phase control for precise modulation. Speed control is achieved by altering armature voltage, field flux, or adding series resistance, while braking is accomplished through regenerative, dynamic, or plugging methods, where the motor acts as a generator to oppose motion.

Operation of DC motor drive is based on following methods:

1. **Power Conversion:** The drive receives an AC input and uses a controlled rectifier to convert it into a controlled DC voltage.

2. **Armature and Field Control:** This DC power is then supplied to the motor's armature, and by controlling the field winding current, the magnetic flux is adjusted.

3. **Torque and Motion:** The interaction between the magnetic field and the current in the armature creates a force (Lorentz force) that generates torque, causing the motor to rotate.

4. **Commutation:** As the motor rotates, the commutator reverses the current direction in the armature windings, ensuring continuous rotation.

Controlling the DC motor drive is based on the following parameters:

1. **Starting:** High initial currents are managed by controlling the resistance in the armature circuit or by using soft starters, allowing for a gradual application of power.

2. **Stopping:** The motor can be stopped using various braking techniques.

   Regenerative Braking: The motor acts as a generator, feeding power back to the source, which is useful when there's a load to absorb the power. Dynamic Braking: The motor's armature is connected to a resistor, allowing the generated energy to be dissipated as heat. Plugging: The armature connections are reversed, creating a negative torque to bring the motor to a stop.

3. **Reversing:** For a DC motor, reversing the polarity of the voltage applied to either the armature or the field winding will reverse the direction of the motor's rotation.

4. **Speed Control:** Speed is controlled by manipulating armature voltage or field flux:

   Armature Voltage Control: This is achieved by adjusting the armature voltage using a controlled rectifier or pulse width modulator (PWM), which varies the voltage supplied to the armature windings.

   Field Flux Control: For shunt motors, the field flux can be controlled by varying the field winding current. Reducing the field current (flux) increases the motor's speed above its rated speed, while increasing the flux reduces speed.

3.4 Issues with fuel cells in electric vehicle application GTUexam-3170923-SUM2025

Main issues with fuel cells in electric vehicles include their high cost and short lifespan compared to conventional engines, challenges with hydrogen storage and the lack of a widespread refueling infrastructure, issues with fuel cell durability such as catalyst poisoning or degradation from flooding/drying, poor performance in dynamic driving conditions (especially for pure fuel cell systems), and challenges with thermal and water management within the vehicle.

Economic & Infrastructure Challenges

- **Cost:**

  Fuel cells, and especially the hydrogen fuel needed, are currently expensive to produce and purchase.

- **Hydrogen Infrastructure:**

  There is a lack of widespread hydrogen refueling stations, which is a significant hurdle for public adoption and convenience.

- **Hydrogen Production & Storage:**

  Producing, transporting, and storing hydrogen is more complex and costly than gasoline or diesel, and storing it on board a vehicle is difficult due to its low density.

Durability & Performance Issues

- **Short Lifespan/Durability:**

  Fuel cells are not as durable as conventional engines, and achieving the desired operational life for automotive applications remains a significant challenge.

- **Faults and Degradation:**

  Fuel cells are prone to various faults, including flooding (too much water), drying (too little water), fuel or oxidant starvation, and catalyst poisoning from impurities like carbon monoxide.

- **Transient Performance:**

  Fuel cells struggle with the high power demands of dynamic driving, such as during acceleration or high-power demands from the vehicle. They also cannot effectively regenerate braking energy like batteries can.

Operational & Engineering Challenges

- **Thermal and Water Management:**

  Effectively managing the heat and water produced by the fuel cell system within the vehicle is a complex engineering challenge.

- **Cold Start Performance:**

  Fuel cells can have difficulty starting in cold conditions, which adds complexity to the system.

Fig 3.3: Fuel cell power transfer unit

To overcome these limitations, many fuel cell vehicles use a hybrid system, incorporating batteries or supercapacitors to handle peak power demands, capture regenerative braking energy, and extend the fuel cell›s lifespan, thereby improving overall durability and performance.

3.5 Power Flow Control in EV Drive

Efficient power flow control is essential in electric drive-trains to ensure optimal performance, energy efficiency, and system safety. This control is achieved through an integrated architecture involving several key subsystems

Motor Controller Unit (MCU): The MCU is responsible for managing motor torque and speed by interpreting accelerator and brake inputs. It governs switching sequences in the inverter and regulates motor behaviour based on driving conditions.

Inverter and Converter: The inverter converts direct current (DC) from the battery into alternating current (AC) for the motor (in AC drive systems), while the DC-DC converter steps down high-voltage battery output to low-voltage auxiliary systems (like lighting, infotainment, and controls).

Battery Management System (BMS): The BMS monitors and manages the battery's state of charge (SOC), voltage, current, temperature, and cell balancing. It ensures safe and efficient energy use and protects against overcharging or deep discharging. Drive Control Algorithms: These algorithms optimize the interaction between vehicle dynamics and powertrain efficiency. They manage tasks such as regenerative braking, traction control, torque vectoring (in multi-motor setups), and adaptive drive modes. Through the collaborative function of these systems, electric vehicles maintain seamless acceleration, regenerative energy recovery, thermal management, and intelligent energy distribution—hallmarks of advanced electric mobility.

3.6 Drive Modes

In electric vehicles (EVs) and hybrid drive systems, power flow is managed through multiple drive modes that adapt to varying driving conditions, road gradients, and the state of charge (SOC) of the battery. The main drive models include EV Mode, Assist Mode, Charge Mode, and Regenerative Mode, each serving specific purposes to optimize energy efficiency and performance. Mode operates using only the battery as the power source. It is primarily used during low-speed city driving, where quiet and emission-free operation is preferred. This mode minimizes fuel use and is ideal for short-distance travel in urban areas. Assist Mode engages both the electric motor and regenerative braking systems. It is typically activated during acceleration and deceleration phases. The electric motor assists the internal combustion engine (if present), reducing fuel consumption, while regenerative braking recovers energy during deceleration, enhancing efficiency. Charge Mode utilizes the engine or regenerative braking to recharge the battery. This mode is beneficial when the SOC is low, and additional battery charging is needed. The internal combustion engine (ICE) may run in optimal efficiency zones to charge the battery, or energy may be captured during braking. Regenerative Mode enables the electric motor to function as a generator, particularly during braking and downhill driving. In this mode, kinetic energy is converted into electrical energy and fed back into the battery, improving overall energy economy.

The selection of these modes is dynamically managed by the vehicle's control system based on driver input, battery SOC, and road gradient, ensuring optimal performance and energy utilization under various driving scenarios. Power flow depends on driver demand, SOC, and road gradient.

3.7 Fuel Efficiency Analysis of Electric Drive-Trains

The fuel efficiency of vehicle propulsion systems varies greatly depending on the technology employed. Internal Combustion Engine (ICE) vehicles, powered by petrol or diesel, typically operate at an efficiency of around 25–35%. A large portion of the fuel's energy is lost as heat and friction, resulting in relatively low mechanical output. In contrast, hybrid systems—which combine an ICE with an electric motor—achieve improved efficiency in the range of 35–50%. This is primarily due to the motor assisting during acceleration and the use of regenerative braking, which recovers some energy that would otherwise be wasted. Fully Electric Vehicles (EVs), however, offer the highest energy efficiency, typically ranging between 85–90%. These systems avoid the inefficiencies of combustion engines and benefit from minimal mechanical losses, efficient electric motors, and effective regenerative braking systems. Several factors influence the efficiency of electric drive-trains. The type of electric motor used (e.g., Permanent Magnet Synchronous Motor or Induction Motor) and its control strategy significantly affect overall performance. Regenerative braking plays a key role, as it allows the system to recover kinetic energy during deceleration, converting it back into electrical energy. Battery efficiency also contributes, with modern lithium-ion batteries offering high round-trip efficiency during charging and discharging cycles. Additionally, a well-designed thermal management system is essential to maintain optimal operating temperatures for the battery and power electronics, thereby minimizing energy losses. Finally, vehicle load and terrain conditions, such as hilly roads or heavy cargo, influence the energy demand and, consequently, the system's operational efficiency.

In summary, while ICE and hybrid vehicles have relatively limited efficiency, electric drive-trains stand out due to their superior energy conversion and intelligent power management.

Electric Propulsion Unit

The electric propulsion system consists of the electric motor, power converter, and control unit.

The electric propulsion unit (EPU) forms the heart of modern electric vehicles (EVs) and hybrid electric vehicles (HEVs), converting electrical energy into mechanical motion while ensuring efficient control and power management. It primarily comprises three core subsystems: the electric motor, the power converter, and the control unit. These components work in perfect synchrony to deliver high performance, smooth operation, and optimal energy efficiency under varying driving conditions.

The electric motor is the prime mover in the propulsion system, responsible for producing the torque required to drive the wheels. Various motor technologies are employed depending on vehicle performance, cost, and efficiency requirements. Permanent Magnet Synchronous Motors (PMSM) are widely used in premium EVs due to their high torque density, superior efficiency, and excellent control over a wide speed range. Induction Motors (IMs) are preferred in some applications (e.g., Tesla's earlier models) due to their robustness, simpler rotor construction, and ability to operate without rare-earth magnets. Switched Reluctance Motors (SRM) are also gaining attention for their high-speed capabilities, fault tolerance, and cost-effectiveness, though they often require advanced control to minimize torque ripple and acoustic noise. Regardless of the motor type, the key design parameters—torque-speed characteristics, efficiency map, and thermal limits—determine the propulsion performance of the EV.

To drive the motor, the power converter acts as the intermediary between the high-voltage battery and the motor windings. Typically implemented as a Voltage Source Inverter (VSI) using IGBT or SiC MOSFET technology, the converter performs DC-to-AC conversion for AC motors or DC-to-DC chopper control for DC motors. Advanced converters incorporate high switching frequencies to reduce harmonic losses and allow compact filter design, while liquid or forced-air cooling systems manage the substantial heat generated during rapid power switching. Furthermore, converters are designed for bidirectional power flow, enabling regenerative braking, where the motor operates as a generator to recover kinetic energy and store it back into the battery or a supercapacitor. This regenerative capability improves overall energy efficiency and extends vehicle range, especially in stop-and-go urban driving.

The control unit, often termed the Motor Control Unit (MCU) or Electric Drive Control Unit (EDCU), acts as the brain of the propulsion system. It regulates the motor's torque and speed based on driver inputs (throttle position), vehicle operating conditions, and higher-level commands from the Vehicle Control Unit (VCU). Advanced control techniques such as Field-Oriented Control (FOC), Direct Torque Control (DTC), or sensor less algorithms are employed to achieve high dynamic response, smooth torque delivery, and minimal losses. The control unit continuously monitors feedback from sensors measuring rotor position, phase currents, DC-link voltage, and temperature to ensure precise actuation and protection against overcurrent, overvoltage, or thermal overload. Integration with thermal management, battery management, and regenerative braking strategies makes the EPU a critical subsystem for achieving desired performance and safety targets.

Beyond basic torque generation, modern EPUs incorporate several advanced features to enhance drivability, efficiency, and reliability. Field-weakening control extends the motor's operational speed beyond the base level by adjusting the stator current vector to counteract back electromotive force (EMF), thereby enabling higher top speeds without compromising safety or efficiency. Torque vectoring capabilities in multi-motor setups allow independent control of wheel torque to improve vehicle handling, traction, and stability under dynamic driving conditions. Furthermore, predictive control algorithms and machine-learning-based estimators are increasingly being deployed to optimize efficiency across varying drive cycles and to adapt motor control strategies to driver behaviour and road conditions.

Another critical aspect is thermal management. Motors, converters, and control electronics generate significant heat, particularly under high-load conditions such as hill climbs, rapid acceleration, or high-speed cruising. Overheating can degrade insulation, reduce efficiency, and shorten the lifespan of components. Therefore, integrated cooling solutions—ranging from finned heat sinks and forced-air fans to advanced liquid cooling with dedicated coolant loops—are used to maintain optimal operating temperatures. Some systems even leverage phase-change materials or oil-spray cooling for ultra-high-performance applications.

From a design perspective, EPUs must strike a balance between power density, efficiency, cost, and reliability. High power density ensures compact packaging and lightweight construction, which directly influences vehicle range and handling. Efficiency improvements—achieved through better motor designs, SiC-based inverters, and optimized control algorithms—reduce energy losses and improve real-world driving range. Reliability and durability are addressed through robust component selection, fail-safe control logic, and redundancy in critical sensing and actuation pathways.

In summary, the electric propulsion unit serves as the primary enabler of electric mobility, delivering precise, efficient, and reliable power conversion from battery to wheels. By integrating advanced motor technologies, high-efficiency converters, and intelligent control algorithms, modern EPUs meet the stringent demands of automotive applications—ranging from urban commuter EVs to high-performance electric sports cars. As innovations in power electronics, control software, and thermal management continue, EPUs will play an even more pivotal role in shaping the future of clean and efficient transportation.

Qualities Required in Electric Motors for Electric Vehicles (EVs) GTUexam-3170923-SUM2025

Electric motors are the primary source of propulsion in Electric Vehicles (EVs) and Hybrid Electric Vehicles (HEVs). They must convert electrical energy from batteries or fuel cells into mechanical energy with high efficiency, reliability, and performance. The motor's design must ensure compactness, quick response, and the ability to handle varying driving conditions such as acceleration, regenerative braking, and load changes.

Electric motors for electric vehicles (EVs) must possess several essential qualities to ensure optimal performance and efficiency. High efficiency is crucial as it ensures maximum conversion of electrical energy into mechanical power, thereby reducing energy losses and extending driving range. Motors must also exhibit high power density and torque density, providing a compact and lightweight design capable of delivering high output per unit mass, which is vital for fast acceleration and efficient space utilization. A wide speed-torque operating range is necessary to deliver high starting torque at low speeds while maintaining constant power at high speeds for highway driving. Additionally, a fast dynamic response is required for immediate torque delivery, ensuring smooth acceleration and deceleration under varying load conditions.

The motor must also support regenerative braking capability, allowing it to function as a generator to recover kinetic energy during braking, thereby improving overall efficiency. Reliability and durability are equally important, as the motor must withstand thermal stresses, mechanical vibrations, and frequent start-stop cycles without performance degradation. A compact size and low-weight design improve the vehicle's power-to-weight ratio and maximize space for passengers or cargo. Furthermore, low noise and vibration levels are essential to enhance passenger comfort and minimize acoustic emissions during operation. Motors should also have a high overload capacity to handle short bursts of high-power demands during overtaking or while climbing steep gradients. Finally, cost-effectiveness and ease of manufacturing are key considerations to ensure economic feasibility for large-scale production while minimizing reliance on expensive rare-earth magnets.

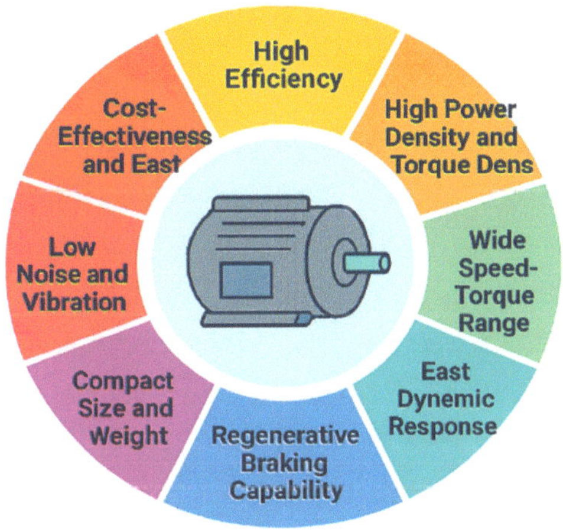

Fig 3.4: Qualities in electric motor

3.8 Torque-speed characteristics in motors GTUexam-3170923-SUM2025

Fig. shows the Torque-speed characteristics of torque–speed characteristics of EV Motors. The torque–speed characteristics of electric vehicle (EV) motors determine their performance, efficiency, and suitability for various driving conditions. Different types of motors display distinct torque–speed profiles, which affect vehicle acceleration, cruising ability, and regenerative braking. In DC Shunt Motor-The torque–speed characteristic of a DC shunt motor shows nearly constant torque over a wide range of speeds up to the rated value. At lower speeds, the torque remains almost flat due to the constant field flux provided by the shunt winding. As speed increases beyond the rated point, armature reaction and increased losses cause a gradual drop in torque. These motors are simple and cost-effective, but they are less efficient than modern AC machines and require frequent maintenance due to the use of brushes and commutators. In Permanent Magnet Synchronous Motor (PMSM)-PMSMs exhibit a flat and high torque output from standstill to the base speed, after which the torque begins to decrease as the motor transitions into the field-weakening region. Since PMSMs operate synchronously with the stator magnetic field, they offer excellent efficiency, high power density, and smooth operation. These features make PMSMs one of the most preferred choices for electric vehicle traction systems, especially in high-performance passenger cars. In Permanent Magnet Brushless DC Motor (PMBLDC)-PMBLDC motors display a nearly constant torque characteristic up to the rated speed, similar to PMSMs, but with slight ripples due to electronic commutation. Beyond the rated speed, torque decreases as the back electromotive force (EMF) rises, limiting current flow. PMBLDC motors are known for high efficiency, compact design, and reduced maintenance compared to DC machines, making them suitable for medium- to high-power EV applications where dynamic performance is essential. Each motor type offers unique advantages for EV applications. DC shunt motors are simpler but less efficient, PMSMs provide superior performance and smooth operation, and PMBLDC motors combine high efficiency with reliability and lower maintenance. Understanding these torque–speed profiles is crucial for selecting the right motor based on vehicle type and performance requirements.

PMSM (specifically Interior PMSM / IPMSM) — best overall choice for mainstream EV traction.

Permanent Magnet Synchronous Motors (PMSMs), particularly the Interior PMSM variant, are widely regarded as the best choice for mainstream electric vehicle (EV) traction due to their exceptional performance characteristics. These motors offer the highest torque density, meaning they can produce more torque per kilogram, allowing for smaller, lighter motors and improved packaging efficiency in vehicles. They also deliver top efficiency across real driving cycles, which directly translates to increased driving range per kilowatt-hour of energy stored.

Furthermore, PMSMs provide a wide constant-power region with excellent field-weakening capability, enabling strong acceleration at low speeds and high-top speeds without requiring additional gearing. Their superior low-speed control and regenerative braking performance ensure smooth torque delivery during vehicle launch, precise traction control, and efficient energy recovery. Additionally, PMSMs benefit from a mature ecosystem of proven inverter technologies, control solutions, manufacturing scalability, and supply chain readiness, making integration into EV platforms more efficient. However, alternative motor types may be preferable in specific cases. Induction Motors (IM) or Switched Reluctance Motors (SRM) are good options when avoiding rare-earth magnets or significantly reducing costs. IMs are rugged, lower-cost, and well-understood, while SRMs are magnet-free, highly robust, but require advanced control strategies and noise, vibration, and harshness (NVH) optimization. For very low-cost, lightweight two-wheelers, Permanent Magnet Brushless DC Motors (PMBLDC) can provide a simpler and more cost-effective solution.

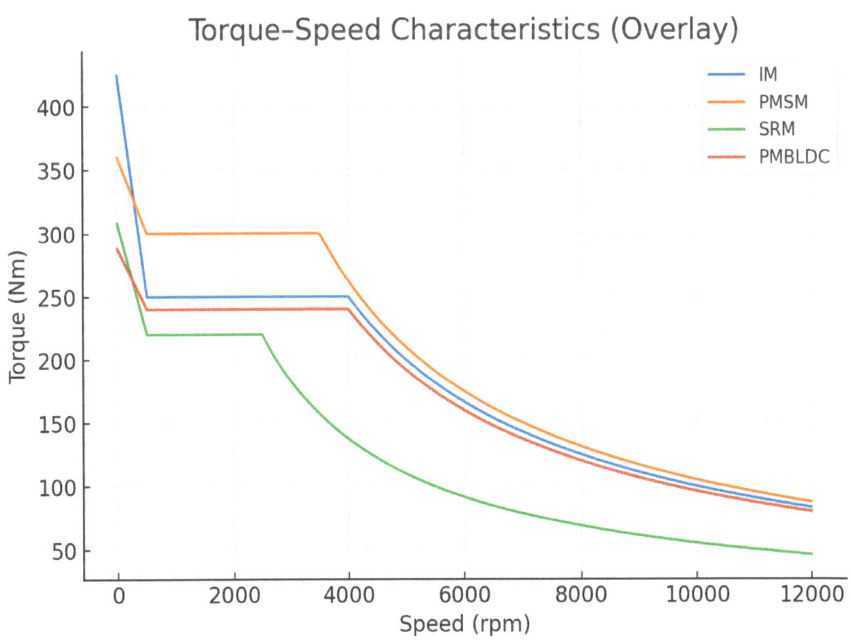

Fig 3.5: Torque -speed characteristics

The curve shows: Induction motor

Induction motors deliver a flat constant-torque output up to their base speed, providing reliable launch performance, followed by a 1/n torque reduction in the field-weakening region. They remain rugged, cost-effective machines, though their light-load efficiency typically trails that of PMSMs, making them common in buses, older EV platforms, and other harsh-duty environments. Permanent-Magnet Synchronous Motors—especially IPMSMs—maintain nearly constant torque to base speed and then extend into a broad constant-power range supported by field-weakening and reluctance torque, offering excellent efficiency, high torque density, smooth control, and strong regenerative capability, which is why they dominate modern EV powertrains. Switched Reluctance Motors provide strong initial torque, but generally exhibit lower average torque, noticeable torque ripple, and acoustic noise, along with a limited high-speed constant-power region unless heavily optimized; nevertheless, their low cost, robustness, and magnet-free architecture make them suitable for cost-sensitive or rare-earth-free traction and auxiliary applications. PMBLDC machines with trapezoidal back-EMF deliver a flat torque region with some commutation ripple and offer moderate field-weakening capability, typically narrower than that of IPMSMs; they are widely used in e-bikes, scooters, light EVs, and general-purpose drives such as fans and pumps, and can be scaled to larger platforms with careful design.

Justification of the DC Shunt motor & Induction motor fundamentals and differences

DC shunt motor and an induction motor are not the same; they are fundamentally different types of electric motors, operating on different types of current (DC vs. AC) and having distinct construction and operating principles, with DC shunt motors using brushes and a commutator for current to the rotor, and induction motors using electromagnetic induction from the stator's magnetic field to create current in a squirrel-cage rotor. Breakdown differences:

DC Shunt Motor

A DC shunt motor is powered by direct current (DC) and is constructed using brushes and a commutator to deliver electrical power to the armature windings on the rotor. These windings are connected in parallel, or shunt, with the field windings on the stator. The motor operates on the principle of converting DC electrical energy into mechanical energy by generating a magnetic field in the stator, which interacts with the magnetic field produced by the rotor, with the commutator ensuring proper current direction. Speed control in a DC shunt motor is inherently variable and can be achieved by adjusting the voltage supplied to the armature and/or the field windings, allowing for smooth and flexible operation across a wide range of speeds.

Induction Motor

An induction motor is powered by alternating current (AC) and is widely used in electric vehicle applications due to its ruggedness and simplicity. It has no electrical connections to the rotor, which consists of a short-circuited "squirrel cage" made of copper or aluminium bars. The operating principle is based on electromagnetic induction, where the stator windings produce a rotating magnetic field from the AC supply, inducing a current in the rotor bars. This induced current generates a magnetic field in the rotor, which interacts with the stator field to produce torque. The motor's speed is primarily determined by the frequency of the AC supply voltage, making frequency control essential for regulating its performance.

Key Differences at a Glance

- **Current Type:**

  DC for a shunt motor, AC for an induction motor.

- **Rotor Power:**

  Direct connection to the rotor via brushes in a DC shunt motor, versus no direct electrical connection in an induction motor.

- **Commutation:**

  DC shunt motors use a commutator to switch the current in the rotor windings, while induction motors do not need a commutator.

- **Maintenance:**

  DC shunt motors generally require more maintenance due to brush wear, whereas induction motors require less maintenance, with only bearings sometimes needing lubrication.

A driving control method for an Induction Motor in an electrical vehicle application GTUexam-3170923-SUM2025

In electric vehicle applications, the induction motor (IM) is a preferred choice due to its robustness, low cost, and reliable performance. However, precise speed and torque control require advanced drive control strategies. The driving control method for IM-based EVs begins with the DC power from the traction battery, which is converted into a variable voltage and variable frequency AC supply through a voltage source inverter controlled by PWM techniques. The driver's torque or speed demand, obtained from accelerator and brake pedals, is processed by a digital controller that generates reference signals for the inverter. Different control strategies are employed to regulate motor performance. The simplest is the constant V/f (voltage-to-frequency) control, which is cost-effective but offers limited dynamic response. More advanced methods include Field Oriented Control (FOC), which decouples torque and flux, providing DC motor–like performance with fast dynamics and efficient regenerative braking, and Direct Torque Control (DTC), which enables rapid torque response with higher ripple. During operation, the controller ensures smooth transitions between acceleration, cruising, hill climbing, and regenerative braking modes by adjusting inverter output and maintaining desired torque. This closed-loop system, supported by current and speed sensors, ensures high efficiency, quick response, and reliable performance of the induction motor drive, making it highly suitable for modern electric vehicle applications. Induction motors (IMs) are widely used in EVs because of their ruggedness, low cost, high efficiency, and maintenance-free operation. But, unlike DC motors, IMs don't have direct torque/speed control. So, power electronics + control strategies are required for precise drive performance.

There are mainly two types of control strategies:

(a) V/f Control (Scalar Control)

- Keeps the ratio of Voltage/Frequency constant.
- Simple, inexpensive, stable.
- Good for constant torque applications.
- Limitation: Poor dynamic response (not suitable for fast-changing driving conditions).

(b) Vector Control (Field Oriented Control – FOC)

- Decouples torque and flux like in DC motors.
- Torque control via q-axis current; Flux control via d-axis current.
- Provides fast dynamic response, high efficiency, regenerative braking, ideal for EV driving.
- Needs current sensors + rotor flux estimation.

(c) Direct Torque Control (DTC)

- Directly controls torque and flux without an inner current loop.
- Faster response than FOC, but higher torque ripple.

Comparison Table of Control Methods in EV IM Drives

Table 3.2: Comparison of control methods in EV

Control Method	Complexity	Dynamic Response	EV Suitability
V/f Control	Low	Moderate	Entry-level EVs
FOC (Vector)	High	Excellent	Most EVs today
DTC	Medium	Very Fast	High-performance EVs

For EV applications, Field Oriented Control (FOC) is most widely used because it gives DC-motor-like torque control, high efficiency, and smooth regenerative braking, making it the industry standard.

Construction and working of the Switched Reluctance Motor GTUexam-3170923-SUM2025

The SRM operates based on the principle of variable reluctance. When a stator phase winding is energized by a DC source via a power electronic converter, a magnetic field is produced. The rotor moves to a position where magnetic reluctance is minimum, i.e., where the stator and rotor poles align. Sequential switching of stator phases creates a rotating magnetic field, causing continuous rotation of the rotor. The absence of windings or magnets in the rotor reduces losses and improves efficiency, making SRMs suitable for electric and hybrid vehicles. The Switched Reluctance Motor (SRM) consists of a stator and a rotor, both having salient poles. The stator is equipped with concentrated windings placed on alternate poles. Rotor is made of laminated steel without windings or permanent magnets, making it robust and simple. Unequal reluctance in the air gap between stator and rotor creates torque when stator windings are energized.

3-ϕ SRM-switched reluctance motor drive GTUexam-3170923-SUM2025

To drive a three-phase switched reluctance motor (SRM), a specialized power converter is required to sequentially energize the stator phase windings with a unipolar current. The most common and flexible circuit for this purpose is the asymmetric half-bridge converter.

Asymmetric half-bridge converter circuit-For a three-phase SRM, the asymmetric half-bridge converter uses two power switches (e.g., IGBTs or MOSFETs) and two diodes for each of the three phases. The circuit is fed by a DC voltage source, often from a rectifier and filter capacitor.

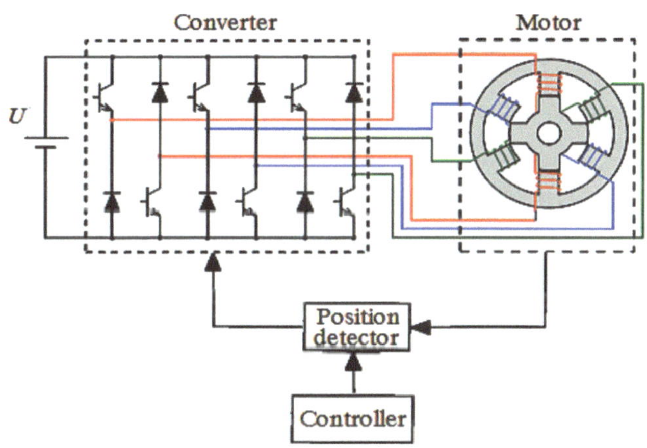

Fig 3.6: 3-ϕ SRM

Why PMSM (Preferably Interior PMSM) is Best for EV Traction?

Permanent Magnet Synchronous Motors (PMSMs), particularly Interior PMSMs (IPM), are widely regarded as the optimal choice for electric vehicle (EV) traction due to their superior performance across critical parameters. They offer high torque density, producing more Newton-meters (Nm) per kilogram, which allows for compact drive units and better vehicle packaging. Their high efficiency, both at peak and across the drive cycle, translates to longer range per kWh and reduced cooling requirements, lowering overall system mass. PMSMs also feature a wide constant-power speed range, enabling the use of single-speed transmissions while maintaining strong mid- and high-speed performance. Additionally, their smooth torque control and excellent regenerative braking capability provide refined noise, vibration, and harshness (NVH) characteristics and precise torque tracking essential for traction control, anti-lock braking, and advanced driver assistance systems (ADAS). IPMs have the added advantage of reluctance torque contribution, which supplements permanent magnet torque, reducing rare-earth magnet mass while sustaining high performance.

However, PMSMs may not be the best choice in applications requiring cost minimization or elimination of rare earth materials, where Induction Motors (IM) or Switched Reluctance Motors (SRM) become viable alternatives. IMs are mature, robust, and magnet-free, while SRMs avoid magnets entirely but require NVH refinement. For lower-cost applications like scooters or e-rickshaws, Permanent Magnet Brushless DC Motors (PMBLDC) are pragmatic due to simpler inverter control.

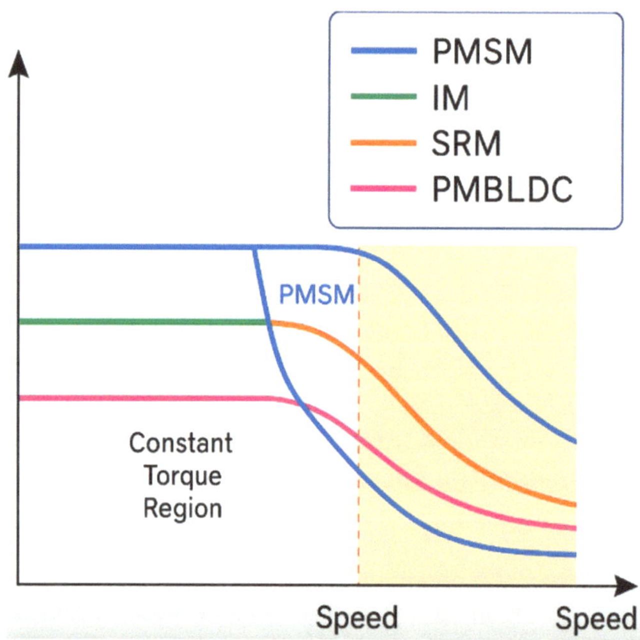

Fig 3.7: Comparison of speed-torque for different types of motors

Numerical Comparison

IM: 250 Nm from 0–4000 rpm → $P \approx 250 \times 4000 \times \frac{2\pi}{60} \approx 104\ \mathrm{kW}$, torque drops as 1/n beyond base speed.

PMSM (IPM): 300 Nm to ~3500 rpm → $P \approx 110\ \mathrm{kW}$ at base; constant-power region extends to 10–12k rpm.

SRM: ~220 Nm to ~2500 rpm → smaller constant-power region; average torque limited by ripple.

PMBLDC: ~240 Nm to ~4000 rpm → reasonable constant-power capability, but commutation limits smoothness.

Torque–Speed Characteristics

Electric vehicle (EV) motors typically operate in two primary regions to balance performance across different speed ranges. In the constant torque region, torque remains nearly flat from standstill up to the motor's base speed, ensuring strong low-speed acceleration and smooth vehicle launch. Beyond this point, the motor transitions into the constant power region, where torque decreases approximately in inverse proportion to speed (1/n). This allows the motor to maintain consistent power output, enabling efficient high-speed performance and extending the vehicle's top-end capability without requiring additional gearing.

Equations:

- **Power:** P = T × ω (W)

  where T = Torque (Nm), ω – Angular Speed (rad/s) ▪ $\omega = \frac{2\pi \times n}{60}$
- **Constant Power Region:** T ∝ 1/n (Torque drops inversely with speed)

Motor Strengths Limitations:

IM Rugged, low cost, wide speed range, slightly lower efficiency

PMSM (IPMSM): High torque density, excellent efficiency, wide constant power range. High cost due to magnets

SRM Magnet-free, robust, simple rotor High noise and torque ripple

PMBLDC Compact, efficient for light EVs, Ripple & narrower constant power range

3.8.1 flux weakening in permanent magnet machines GTUexam-3170923-SUM2025

Flux weakening is a motor control strategy used in permanent magnet synchronous motors (PMSMs) for high-speed operation, where stator currents are intentionally used to generate a magnetic flux that opposes and weakens the permanent magnets› flux. This technique allows the motor to maintain constant output power by effectively extending the operating speed beyond the base speed, compensating for the increasing back EMF at higher speeds, and keeping the motor within the inverter's voltage and current limits.

Purpose of Flux Weakening:By extending operating speed, at higher speeds, the back electromotive force (back EMF) in the motor increases, requiring a higher terminal voltage to maintain torque. Flux weakening allows the motor to operate above its base speed by reducing the effective air-gap flux, which is proportional to the back EMF, thus extending the motor's speed range. Explaining the achieve constant power operation, here by weakening the flux, the motor can continue to operate at a constant power output even as the speed increases beyond the point where maximum torque is available. In flux weakening control of permanent magnet machines, the stator current is regulated with a specific phase angle so that the resulting stator flux opposes the flux produced by the rotor permanent magnets. This opposing flux creates a demagnetizing effect, which reduces the overall air-gap flux and enables the motor to operate at higher speeds, even under the constant voltage and current limitations imposed by the power inverter. To achieve this, advanced control strategies such as direct flux vector control and vector current control are employed, where the stator current components are precisely managed in a synchronous rotating reference frame to balance torque production with flux weakening.

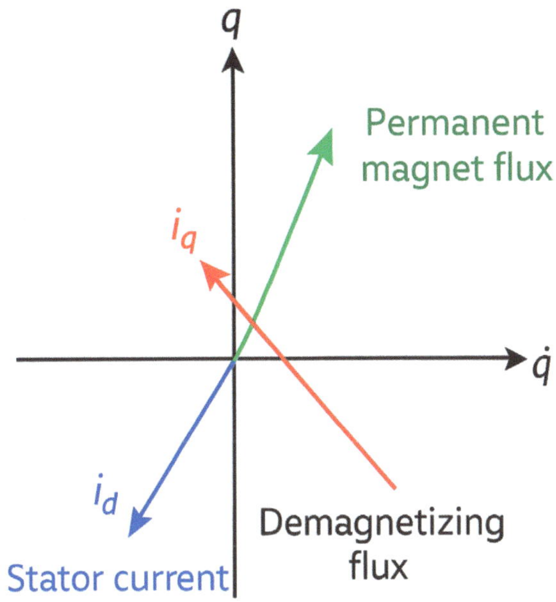

Fig 3.8: Flux weakening control of permanent magnet machines

Electric Components Used in EVs

Electric Vehicles (EVs) rely on several key components to ensure efficient power delivery, smooth operation, and optimal performance. The battery, commonly lithium-ion (Li-ion) or lithium iron phosphate (LFP), serves as the primary energy storage unit, supplying the necessary electrical power to drive the vehicle. The inverter converts the battery's direct current (DC) into alternating current (AC), which is required by the electric motor for propulsion. A DC-DC converter is used to step up or step-down voltage levels, ensuring that different vehicle systems, such as auxiliary electronics, receive the correct power supply. The on-board charger facilitates charging by converting alternating current (AC) from the power grid into direct current (DC) suitable for storing in the battery. The controller plays a crucial role in managing motor speed, torque, and overall energy flow to achieve desired driving performance. Additionally, a thermal management system is employed to regulate the temperature of the battery, motor, and power electronics, preventing overheating and ensuring safe, efficient operation.

Configuration & Control of Electric Motors

Electric motors used in Electric Vehicles (EVs) are designed to convert electrical energy into mechanical motion efficiently. Their configuration and control determine vehicle performance, efficiency, and drivability. Proper control ensures optimal torque production, speed regulation, and smooth power delivery while maintaining energy efficiency. DC motor drives are one of the oldest and simplest motor technologies used in EVs.

They can be classified into two primary types:

3.8.1 Brushed DC Motors (BDC)

These motors use mechanical brushes to supply current to the rotating armature. They are cost-effective and provide high starting torque but require regular maintenance due to brush wear. A brushed DC (BDC) motor is an electric motor that converts direct current (DC) into mechanical energy by using a set of stationary brushes to transfer power to a rotating commutator. These are one of the most basic and cost-effective motor types, known for their simple control and high starting torque.

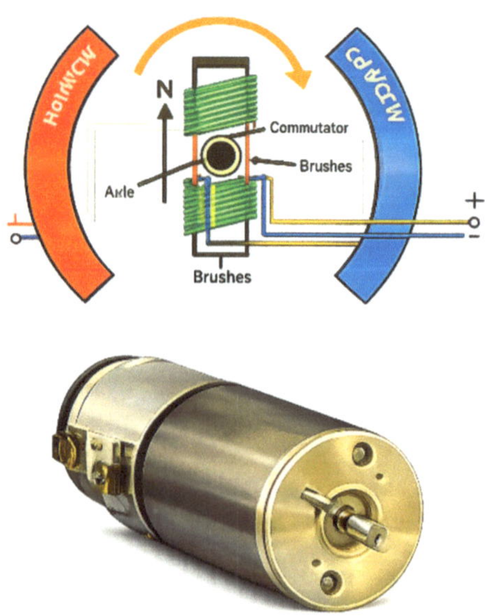

Fig 3.9: Brushed DC motor

The working principle of a BDC motor brushed DC motor, consists of four main components:

Stator: The stationary outer part of the motor that provides a constant magnetic field. In smaller motors, this field comes from permanent magnets, while in larger motors, it is created by electromagnets called field windings.

Rotor (or Armature): The rotating inner part of the motor that contains one or more wire windings.

Commutator: A segmented conductive copper sleeve mounted on the rotor's axle. As the rotor spins, the commutator segments come into contact with the brushes.

Brushes: Stationary contacts, typically made of carbon, that maintain contact with the commutator. They deliver current from the power source to the rotor windings.

The process of operation is as follows:

DC voltage is applied across the brushes, which passes current to the rotor's windings through the commutator. This current creates a magnetic field in the rotor, causing it to act as an electromagnet. The rotor's magnetic poles are repelled by the stator's magnetic poles, causing the rotor to turn. As the rotor spins, the brushes slide onto the next segments of the commutator, reversing the direction of the current in the rotor's windings. This reversal of current flips the rotor's magnetic poles, causing the continuous attraction and repulsion that keeps the motor spinning.

Types of brushed DC motors-BDC motors can be categorized by the construction of their stator field winding: Permanent Magnet (PM): Uses permanent magnets for the stator's magnetic field. This is the most common and simplest type, often found in small devices and toys. Series-Wound: The field winding is connected in series with the armature coil. This provides high starting torque and is used for heavy-duty applications like cranes and winches. Shunt-Wound: The field winding is connected in parallel (shunted) to the armature coil. This results in excellent speed control and is ideal for constant-speed applications such as fans and pumps. Compound-Wound: A combination of series and shunt windings, offering a mix of both characteristics.

3.8.2 Brushless DC Motors (BLDC)

These motors eliminate brushes by using electronic commutation, which improves efficiency, reduces maintenance, and increases lifespan. Advantages and disadvantages of BLDC

Table 3.3: Advantages & disadvantages of BLDC

Advantages	Disadvantages
Low cost due to a simpler design and mature manufacturing processes.	Wear and tear on the brushes and commutator require periodic maintenance or replacement.
Simple control, as speed is generally proportional to the supply voltage.	Lower efficiency compared to brushless motors due to friction and electrical noise.
High starting torque, making them suitable for applications that need to accelerate quickly.	High electrical noise (EMI) is generated from the arcing between the brushes and commutator.
Robust and reliable in many operating environments.	Shorter lifespan, typically 2,000 to 5,000 hours, compared to brushless motors.

3.8.3 Driver circuit for 3 ϕ BLDC motor GTUexam-3170923-SUM2025

A 3-phase Brushless DC (BLDC) motor driver circuit controls the motor's speed and torque by electronically commutating the stator windings. Instead of mechanical brushes, it uses a three-phase inverter bridge with switching devices like MOSFETs or IGBTs. A microcontroller or driver IC receives rotor position feedback from Hall effect sensors (or estimates it using sensorless methods) to generate the correct Pulse Width Modulation (PWM) signals for the inverter.

Explanation of the components: Microcontroller (MCU): This acts as the "brain" of the driver. It executes a control algorithm based on rotor position feedback to determine the sequence and timing of the switching signals for the power stage. Position Sensors (Hall Effect Sensors): A BLDC motor requires an electronic controller to know the rotor's position to properly energize the stator windings. For this purpose, three Hall effect sensors are typically embedded inside the motor, spaced 120 electrical degrees apart. They provide feedback to the MCU, signaling when the rotor's magnetic poles are aligned with the stator windings. Gate Driver ICs: These integrated circuits amplify the low-power control signals from the microcontroller to provide the higher-voltage and higher-current signals required to rapidly switch the power MOSFETs. They prevent high voltage and current from damaging the MCU and also prevent "shoot-through," a condition where both the high-side and low-side transistors in a half-bridge are turned on simultaneously.Three-Phase Inverter Bridge: The core of the power stage is a three-phase, six-transistor inverter bridge, usually made of N-channel MOSFETs or IGBTs. It consists of three "half-bridges," with each half-bridge driving one of the motor's three phases.

Connections & set up of IGBTs with 3 phase supply are like:

- **High-Side Switches (Q1, Q2, Q3):** Connected to the positive DC power rail.
- **Low-Side Switches (Q4, Q5, Q6):** Connected to the ground.
- **Phase Connections (U, V, W):** The outputs are connected to the three windings of the BLDC motor.

  Current and Voltage Sensors: These sensors provide feedback to the MCU for closed-loop control.
- **Current Sensing:** Detects the current flowing through the motor phases to implement over-current protection and torque control.

- **Voltage Sensing:** Monitors the DC bus voltage and detects the motor's back-EMF in sensorless control systems.

  DC Power Supply: Provides the necessary direct current to the driver circuit. A capacitor is often included to smooth the voltage supply and provide current during rapid switching.

Operation of 3 ϕ BLDC motor:

Rotor Position Detection: The Hall effect sensors continuously monitor the rotor's magnetic field and send a 3-bit digital signal to the microcontroller. Each of the six possible combinations of this signal corresponds to a specific 60° electrical position of the rotor.

Electronic Commutation: Based on the Hall sensor feedback, the microcontroller determines which two of the motor's three windings need to be energized next. It sends control signals to the gate drivers to switch on the appropriate pair of MOSFETs in the inverter bridge. For example, in a six-step trapezoidal control, a sequence might energize Phase A and Phase C, leaving Phase B floating.

Generating the Rotating Field: By sequentially switching the pairs of windings, the controller creates a rotating magnetic field in the stator. The rotor's permanent magnets try to align with this field, causing the motor to rotate.

Speed Control (PWM): The speed of the motor is controlled by adjusting the PWM duty cycle of the signals sent to the MOSFETs. A higher duty cycle results in a higher average voltage supplied to the windings, increasing the motor speed.

Feedback and Control Loop: In a closed-loop system, the MCU reads sensor data, such as speed or current, and adjusts the PWM duty cycle in real-time to maintain the desired speed under varying loads.

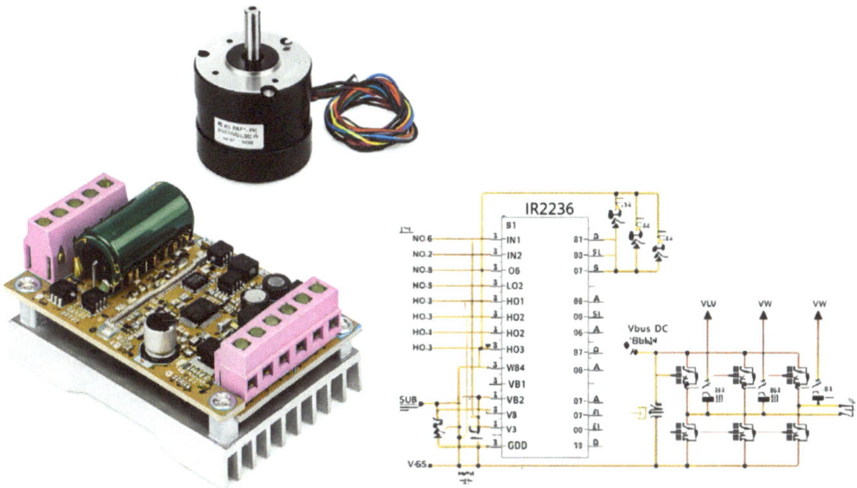

Fig 3.10: 3 ϕ BLDC motor driver circuit

Brushed DC (BDC) motors remain relevant in various fields despite the increasing use of brushless technologies. Their simplicity, low manufacturing cost, and reliable performance in short duty cycles make them ideal for applications where long operational life and high efficiency are not primary requirements. In the automotive sector, BDC motors are commonly used in automatic windows, seat adjustment mechanisms, and windshield wipers, where intermittent operation and cost-effectiveness are key factors. In consumer electronics, they are frequently found in low-cost toys, small household appliances, and vacuum cleaners due to their ease of integration and affordability. The industrial sector also employs these motors in equipment such as cranes, paper machines, and steel rolling mills, where high starting torque is beneficial for handling heavy loads. Additionally, in power tools like drills and saws, BDC motors provide the necessary torque for demanding tasks while keeping production costs manageable.

Control

The speed and torque of DC motors are controlled by varying the armature voltage or the field current. Modern controllers use Pulse Width Modulation (PWM) techniques to achieve precise voltage regulation, resulting in smooth acceleration and efficient energy usage.

3.8.4 Features of Brushed DC Motors

Brushed DC (BDC) motors possess several defining features that make them suitable for a wide range of applications, despite the growing use of advanced motor technologies such as Brushless DC (BLDC) motors and AC induction motors. Their primary attributes—simple control structure, high torque generation at low speeds, and good efficiency for low to medium-speed applications—contribute to their ongoing relevance in automotive systems, industrial machinery, consumer electronics, and power tools.

One of the most significant advantages of BDC motors is their simple control structure. Unlike AC motor drives, which require complex electronic controllers and sophisticated algorithms for managing torque and speed, BDC motors can be controlled with basic voltage and current variations. The ability to regulate speed by adjusting the armature voltage or field current makes these motors highly user-friendly and cost-efficient in design. This simplicity reduces the need for advanced circuitry and programming, leading to lower manufacturing and operational costs. Furthermore, the direct commutation through mechanical brushes provides a straightforward approach to energy transfer, eliminating the need for position sensors or intricate feedback systems in basic applications. As a result, BDC motors are widely used in applications where simplicity and low cost are prioritized over long-term durability and efficiency.

Another key feature of BDC motors is their ability to generate high torque at low speeds, making them highly effective for starting and stop-and-go conditions. This characteristic is particularly advantageous in automotive applications, such as windshield wipers, power windows, and seat adjustment systems, where frequent starts and stops are required. The mechanical design of BDC motors allows them to deliver high initial torque without the need for elaborate control systems or additional components, such as gear reducers, in many cases. This makes them ideal for machinery and tools that demand immediate response and powerful motion during start-up phases. Industrial applications such as cranes, paper machines, and rolling mills also benefit significantly from this feature, as these systems often operate under varying loads and require strong initial torque to overcome static resistance.

In addition to simple control and high torque capabilities, BDC motors are known for their good efficiency in low to medium-speed applications. While they may not match the energy efficiency of BLDC or AC induction motors at higher speeds, their performance at lower operational ranges remains competitive. This makes them well-suited for devices and systems that do not operate continuously at high speeds or require extended periods of precision control. For example, power tools like drills and saws, which operate in short bursts, capitalize on this efficiency characteristic. Similarly, consumer electronic devices such as toys and small household appliances benefit from the energy savings offered by BDC motors in low-demand scenarios. The efficiency, while not optimal across all ranges, is sufficient for applications where energy consumption is less critical than reliability and cost.

Moreover, the combination of these features—simplicity, torque performance, and efficiency—creates a motor type that is versatile and adaptable to a variety of tasks. In the context of automotive design, these motors meet performance requirements without significantly increasing production costs, making them ideal for features that enhance convenience and comfort. In industrial settings, their torque characteristics allow them to handle demanding mechanical loads effectively, while their control simplicity ensures minimal downtime and maintenance complexity. Similarly, in consumer and home appliances, where affordability and basic functionality are key, BDC motors provide a practical solution.

The trade-off, however, lies in their mechanical design, particularly the use of brushes, which are subject to wear and tear over time. This introduces the need for periodic maintenance to replace worn-out brushes and maintain performance levels. Nevertheless, in applications where long-term operational life and minimal maintenance are not critical requirements, the benefits of BDC motors often outweigh this limitation.

In summary, BDC motors continue to maintain their place in modern engineering due to their straightforward control requirements, strong low-speed torque performance, and satisfactory efficiency in low to medium-speed ranges. Their versatility enables their use across various industries, from automotive systems and industrial machinery to consumer electronics and power tools. While advancements in motor technology are gradually replacing BDC motors in applications demanding higher efficiency and durability, their inherent features ensure they remain a practical choice in cost-sensitive and moderate-performance environments.

Brushed DC (BDC) motors have been commonly used in e-bikes, early electric vehicles (EVs), and scooters due to their simplicity and cost-effectiveness. In e-bikes and scooters, these motors provide reliable torque for quick acceleration and stop-and-go riding conditions, making them ideal for urban commuting. Early EVs also utilized BDC motors because of their straightforward control system and relatively low manufacturing costs, which helped in the initial development of electric mobility solutions. While modern EVs increasingly rely on advanced motor technologies such as BLDC and AC induction motors, BDC motors remain a practical choice for small-scale vehicles and budget-friendly designs where efficiency and long-term durability are secondary to affordability and ease of control.

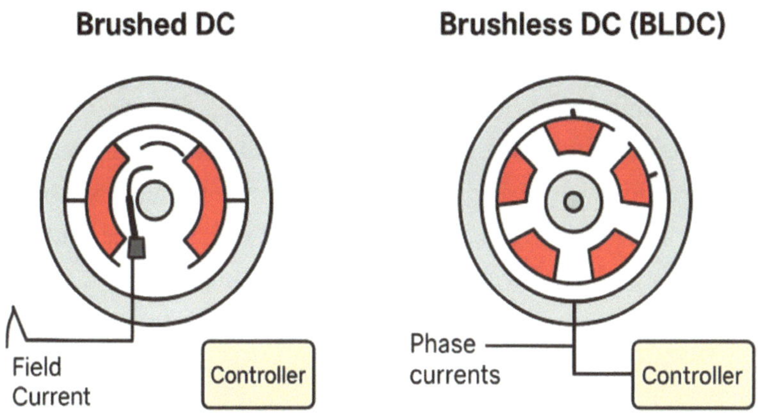

Fig 3.11: BDC, BLDC motors

Induction Motor GTUexam-3170923-SUM2025

Induction Motor (IM) drives are one of the earliest and most widely adopted technologies for electric vehicle propulsion. They can be controlled using either scalar control (V/f control) for simpler, cost-effective applications or vector control for high-performance, dynamic torque and speed response. Unlike permanent magnet machines, induction motors contain no brushes or permanent magnets, which makes them inherently robust, reliable, and comparatively low in cost. Their simple construction and ability to withstand harsh operating conditions further enhance durability. These attributes, combined with the absence of rare-earth materials, make IMs an attractive choice for manufacturers seeking cost efficiency and reduced dependency on critical raw materials. A notable example is the Tesla Model, which employs three-phase AC induction motors, demonstrating their capability to deliver high performance in premium electric vehicles.

3.8.5 Permanent Magnet Motor Drives

Permanent Magnet Motor Drives, including Permanent Magnet Synchronous Motors (PMSMs) and Brushless DC Motors (BLDCs), are among the most popular choices for modern electric vehicle applications. They are characterized by high efficiency, often exceeding 90%, which directly contributes to extended driving range and reduced energy consumption. Owing to their high-power density, these motors are compact and lightweight, making them suitable for efficient vehicle packaging and performance optimization. PMSM and BLDC drives are widely deployed in mainstream EVs; for instance, the Tesla Model 3 rear motor and the Hyundai Kona EV both utilize such technologies. From a control perspective, they rely on advanced strategies such as Field-Oriented Control (FOC) for precise torque and flux regulation, as well as sinusoidal or trapezoidal commutation depending on the motor type. These features make permanent magnet motor drives a leading solution for delivering smooth operation, high performance, and superior energy efficiency in electric mobility.

3.8.6 Switched Reluctance Motors (SRMs)

Switched Reluctance Motors (SRMs) are a distinctive class of electric drives that operate without magnets or windings on the rotor, resulting in a mechanically simple and highly robust construction. Their design offers significant advantages such as low manufacturing cost and excellent tolerance to high operating temperatures, making them suitable for demanding environments. However, a major challenge associated with SRMs is the presence of torque ripple, which requires sophisticated control strategies to achieve smooth performance. Despite this drawback, their cost-effectiveness and thermal resilience make them a practical option in certain electric vehicle applications. For example, the Mahindra e2o has incorporated SRM technology in specific high-temperature use cases, highlighting their potential in niche EV applications where durability and cost considerations are prioritized.

3.8.7 Drive System Efficiency Calculation

The efficiency of an electric drive system is shaped by how smoothly its core components work together. Overall efficiency, η drive), is obtained by multiplying the efficiencies of the motor, inverter, battery, and gearbox.

Because these values compound, small losses accumulate quickly, reducing total performance. For instance, even if each component operates near 95%, the final efficiency drops below that due to the multiplicative effect. This reality makes it essential to fine-tune every subsystem to boost energy use and extend vehicle range

3.8.8 Evaluating Electric Drive Efficiency

The overall drive system efficiency, denoted as η drive, is determined by the combined performance of its major components. It is calculated as the product of the efficiencies of the motor, inverter, battery, and gearbox. Mathematically, this relationship is expressed as

$$\eta \text{ drive} = \eta \text{ motor} \times \eta \text{ inverter} \times \eta \text{ battery} \times \eta \text{ gearbox}.$$

This formulation highlights that even small losses in individual components can accumulate, significantly affecting the overall system efficiency. For example, if the motor, inverter, battery, and gearbox each operate at efficiencies of around 95%, the resulting drive efficiency will be lower than 95% due to the multiplicative effect. Hence, optimizing every element of the drive system is crucial to maximizing the energy utilization and driving range of an electric vehicle.

given

So, the drive system efficiency is usually expressed as:

$$\eta\_\text{drive} = \eta_m \times \eta_i \times \eta b \times \eta g$$

Component	**Symbol**	**Efficiency Range**
Electric Motor	η_m	0.90 – 0.95
Inverter	η_i	0.95 – 0.98
Battery	ηb	0.90 – 0.95
Gearbox/Transmission	ηg	0.96 – 0.98 *(assumed)*

Efficiency Example Calculation

$$\eta\_drive = \eta_m \times \eta_i \times \eta b \times \eta g$$
$$= 0.807 \approx 80.7\%$$

That means about 81% of the energy from the battery reaches the wheels as mechanical power.

Efficiency Range (Estimation):

Minimum	0.90 × 0.95 × 0.90 × 0.96	≈ 73.4%
Maximum	0.95 × 0.98 × 0.95 × 0.98	≈ 86.7%
Typical	0.92 × 0.97 × 0.93 × 0.97	≈ 80.7%

Rememberance

Topic	Highlights
Electric Traction	Motor-driven wheels, replaces ICE
Topologies	Central, In-wheel, Hub
Power Flow	Controlled via MCU, BMS, Inverter
Motors	DC, Induction, PM, SRM
Efficiency	Highest in PM & induction drives

Multiple Choice Questions

1. **Which motor offers the widest constant-power region?**

A) IM

B) PMSM

C) SRM

D) PMBLDC

Answer: B) PMSM

Permanent Magnet Synchronous Motors, especially Interior PMSMs, offer the widest constant-power region due to strong field-weakening capability.

2. **In the constant-power region, torque varies approximately as:**

A) n

B) 1/n

C) n^2

D) Constant

Answer: B) 1/n

- In the constant-power region, torque drops approximately inversely with speed.

3. **Which motor type is magnet-free and exhibits high torque ripple?**

A) IM

B) PMSM

C) SRM

D) PMBLDC

Answer: C) SRM

- Switched Reluctance Motors are magnet-free and exhibit relatively high torque ripple and noise compared to other types

India-GTU Questions

Question	Reference	Marks
Give the operation and controlling of DC motor drive.	chapt3pg72 SUM2025	**03**
What type of operation is known as flux weakening in permanent magnet machines?	chapt3pg94	**03**
Draw torque-speed characteristics of following motors and suggest best for EV with reason. (1) Induction Motor (IM) (2) Permanent Magnet Synchronous Motor (PMSM) (3)	chapt3pg85	**03**
Switched Reluctance Motor (SRM) (4)	chapt3pg91	**03**
Permanent Magnet Brushless DC Motor (PMBLDC)	chapt3pg84	**04**
Prepare and explain a complete driving control method for Induction Motor in an electrical vehicle application.	chapt3pg88	**07**
What are the main issues with fuel cells in electric vehicle application?	chapt3pg74 SUM2025	**03**
Draw and explain the driver circuit for 3 phase SRM-switched reluctance motor drive.	chapt3pg91	
Draw and explain driver circuit for 3 phase BLDC motor.	chapt3pg98,101	**03**
With the help of a neat block diagram explain different subsystems of electric drive train.	chapt3pg62,79 WIN2024	**04**
List the various components of HEV drive train.	chapt3pg64	**04**
Classify the electric motors drives for EV and HEV application.	chapt3pg70	**07**

List the various components of HEV drive train.	chapt3pg64	**07**
Which types of qualities are required in electric motors for EV application?	chapt3pg82	**03**
Draw only torque-speed characteristics of following motors. (1) DC Shunt Motor (2) Permanent Magnet Synchronous Motor (PMSM) (3) Permanent Magnet Brushless DC Motor (PMBLDC).	chapt357pg84	**04**
Briefly explain construction and working of Switched Reluctance Motor.	Chapt3pg 60	**04**
Mention the advantages of hybrid vehicles.	Chapt3pg66	**03**
Give the Comparison of Lithium Ion and Lead Acid Battery Configurations in EV.	chapt3pg66,68 SUM2025	**07**
Mention the advantages of hybrid vehicles.	Chapt3pg66	**03**

Chapter 04

Types of Storage Systems in EV & HEV

4 TYPES OF STORAGE SYSTEMS IN EV & HEV

Electric Vehicles (EVs) and Hybrid Electric Vehicles (HEVs) rely heavily on efficient and robust Energy Storage Systems (ESS). These systems play a pivotal role in ensuring vehicle performance, reliability, and efficiency.

4.1 Energy Storage System requirement GTUexam-3170923-SUM2025

To meet the dynamic demands of modern transportation, the ESS must fulfil several key requirements, such as High Energy Density, High Power Density, Regenerative Braking Compatibility, Compactness and Lightweight Design, Durability, and Longevity. In high-energy-density storage systems, the storage system should be capable of storing large amounts of energy to enable extended driving ranges between charges. This is especially critical for battery electric vehicles (BEVs), where range anxiety is a primary concern. In high Power Density for effective vehicle performance—particularly during acceleration and hill climbing—the ESS must deliver bursts of high power over short durations without compromising system integrity. In regenerative Bra

Battery Performance Parameters in Electric Vehicles-In Electric Vehicles (EVs) and Hybrid Electric Vehicles (HEVs), the performance and efficiency of the energy storage system are determined by several key parameters. These parameters define the capacity, power capability, efficiency, and operational lifespan of the battery. There are all battery performance and characteristics terms explained in detail:

Charge Capacity: Charge capacity refers to the total amount of electric charge a battery can store. It is expressed in ampere-hours (Ah) or milliampere-hours (mAh). For instance, a 100 Ah battery can theoretically deliver 100 A for one hour or 10 A for ten hours before being fully discharged.

Specific Energy: Specific energy represents the amount of energy a battery can store per unit mass and is measured in watt-hours per kilogram (Wh/kg). It indicates how much energy can be stored for a given battery weight. Higher specific energy means a lighter battery for the same energy output, which improves vehicle range.

Energy Density: Energy density is the amount of energy stored per unit volume, measured in watt-hours per liter (Wh/L). It determines how compact a battery can be while still storing a significant amount of energy. High energy density is vital for EVs, where space is limited.

Specific Power-Specific power defines the maximum rate at which energy can be delivered per unit mass, measured in watts per kilogram (W/kg). A higher specific power enables quick acceleration and better performance in high-power demand scenarios such as hill climbing or overtaking.

Charge Efficiency-Charge efficiency is the ratio of energy stored in the battery to the total energy supplied during charging. It indicates how effectively the battery stores the incoming electrical energy. The formula is:

$$\eta_c = \frac{E_{stored}}{E_{supplied}} \times 100$$

Energy Efficiency-Energy efficiency represents the ratio of energy delivered during discharge to the total energy supplied during charging:

$$\eta_e = \frac{E_{discharge}}{E_{charge}} \times 100$$

High energy efficiency reduces energy loss and increases vehicle range.

Life Cycle of Battery-The life cycle refers to the total number of charge-discharge cycles a battery can undergo before its capacity drops to around 80% of its original value. A longer life cycle reduces replacement costs and improves EV sustainability.

An effective ESS must support regenerative braking, which captures and stores kinetic energy during deceleration. This feature not only improves overall energy efficiency but also extends the operational life of the braking system. In Compactness and Lightweight Design The ESS should be designed to occupy minimal space and contribute minimally to the vehicle's weight. A lower weight translates to improved vehicle dynamics and energy efficiency. In durability and Longevity given the harsh operating conditions—temperature fluctuations, vibration, frequent charge-discharge cycles—the ESS must exhibit a long lifecycle with minimal degradation over time. Energy Storage System (ESS) is the Heart of the Powertrain.It serves as both the energy reservoir and the dynamic supplier of power, enabling smooth transitions between various drive modes and supporting hybrid system operations.

Ambient temperature- refers to the surrounding environmental temperature in which a battery operates, and it plays a crucial role in determining the overall performance of an energy storage system. Since battery electrochemistry is highly temperature-dependent, variations in ambient temperature directly influence efficiency, available capacity, and safety. In colder conditions, chemical reactions slow down, leading to reduced energy output and lower driving range, while in excessively hot conditions, the rate of degradation increases, shortening the battery's lifespan and potentially raising safety concerns. Hence, maintaining an optimal ambient temperature is vital for reliable operation and extended life of electric vehicle batteries.

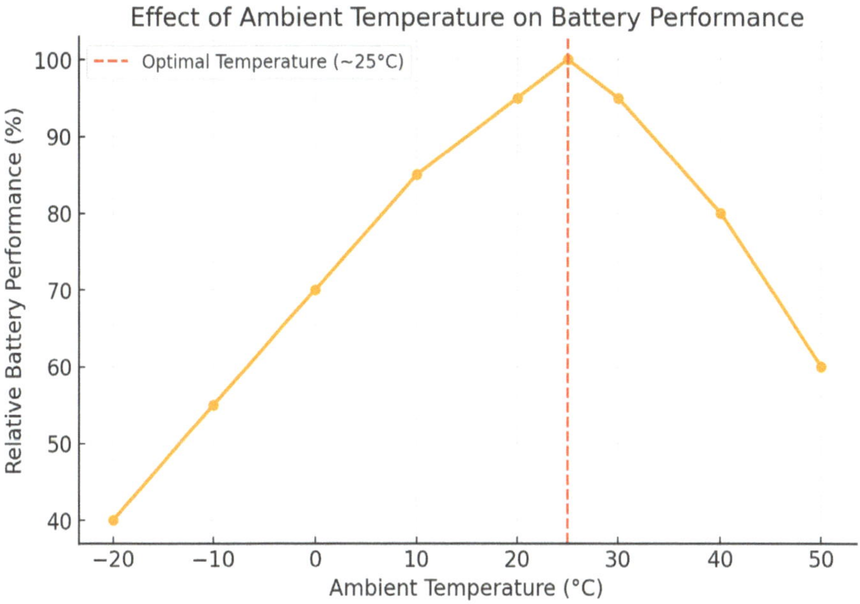

Fig.4.1: Effect of ambient temperature on battery performance

Energy stored- in a battery refers to the total amount of electrical energy that the battery can supply to a load. It is calculated as the product of the battery voltage (V) and its charge capacity (Q), expressed as $E = V \times Q$. Here, V is the nominal voltage of the battery, while Q is the charge capacity measured in ampere-hours (Ah). The resulting energy is typically expressed in watt-hours (Wh) or Kilowatt-hours (KWh). This parameter directly indicates the driving range of an electric vehicle, since a higher stored energy means the vehicle can operate longer before requiring recharging.

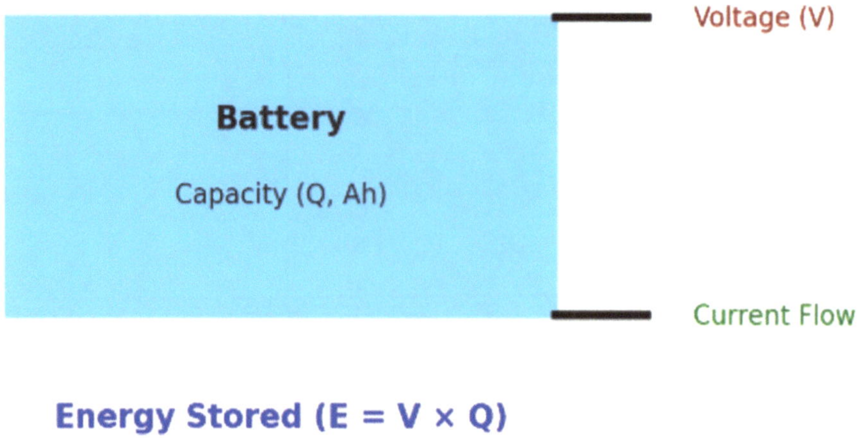

Fig.4.2: Battery storage

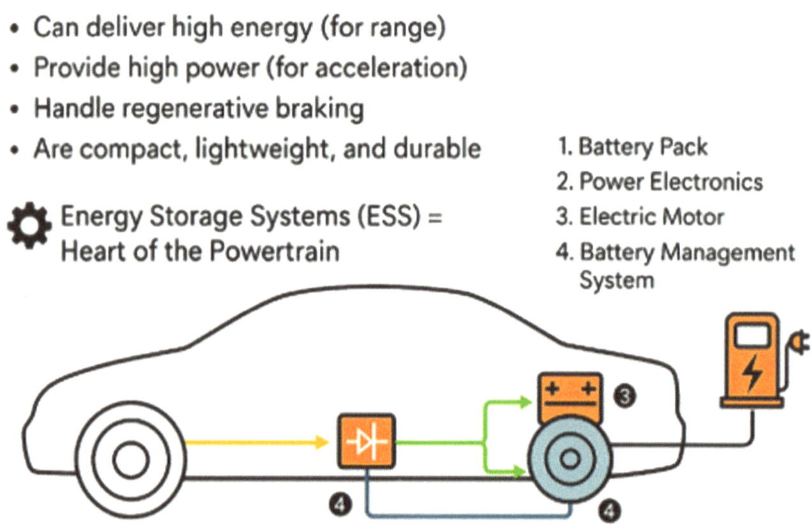

Fig.4.3: Energy Storage System (ESS)

Basic Principle of Supercapacitors-Based Energy Storage System in Hybrid Electric Vehicles GTUexam-3170923-WIN2024

A supercapacitor, also known as an ultracapacitor, is an energy storage device that bridges the gap between conventional capacitors and batteries. It stores energy through electrostatic charge separation rather than electrochemical reactions, resulting in high power density, rapid charging, and long cycle life. The working principle is based on Electrostatic Energy Storage Supercapacitors store energy by accumulating opposite charges on two electrodes separated by an electrolyte and a dielectric or separator. Unlike batteries, no chemical reaction takes place during charging or discharging, leading to fast energy transfer.

$$E = \frac{1}{2}CV^2$$

where **E** is energy stored (Joules), **C** is capacitance (Farads), and **V** is voltage (Volts).

The role of this energy system in Hybrid Electric Vehicles (HEVs) is to provide quick bursts of power during acceleration, absorb regenerative braking energy efficiently, and work in combination with batteries to improve overall efficiency and extend battery life.

The difference between an ultracapacitor and a battery as an energy storage device for EV is explained below GTUexam-3170923-SUM2025

In Electric Vehicles, both batteries and ultracapacitors serve as energy storage devices, but their characteristics are quite different. A battery stores energy through electrochemical reactions, providing high energy density (100–250 wh/kg for Li-ion) and enabling long driving ranges. However, batteries have moderate power density, slower charge–discharge times, limited cycle life, and are more suitable for supplying steady energy during cruising and acceleration. In contrast, an ultracapacitor stores energy electrostatically at the electrode–electrolyte interface, offering very high-power density (10–20 kW/kg), extremely fast charging and discharging, and long cycle life (up to a million cycles). Its drawback is very low energy density (5–10 Wh/kg), making it unsuitable for long-term storage but excellent for short bursts of high power, such as regenerative braking and sudden acceleration. Thus, in EVs, batteries ensure range, while ultracapacitors handle peak power demands, and when combined, they create an efficient hybrid storage system. Sometimes ultracapacitor is denoted as a supercapacitor in vehicle technology.

Table 4.1: Difference between Battery & Ultracapacitor

Parameter	**Battery**	**Ultracapacitor (Supercapacitor)**
Energy Storage Principle	Electrochemical reaction	Electrostatic field (double-layer capacitance)
Energy Density	High (100–250 Wh/kg)	Very low (5–10 wh/kg)
Power Density	Moderate (1–2 kW/kg)	Very high (10–20 kW/kg)
Charge/ Discharge Time	Minutes to hours	Seconds
Cycle Life	1,000–5,000 cycles (Li-ion)	Up to 1,000,000 cycles
Efficiency	70–90%	>95%
Self-discharge	Low	High
Best Use in EVs	Long driving range, main energy source	Short bursts of power, regenerative braking, load levelling
Merits	High energy, good for cruising	Very high power, rapid charging, long life

Battery Management System (BMS):

The Battery Management System (BMS) is a critical component of electric and hybrid vehicles, responsible for ensuring the safe, efficient, and reliable operation of the battery pack. It continuously monitors key parameters such as the state of charge (SOC), cell voltage, current flow, and temperature across all cells and modules within the battery. By doing so, it prevents overcharging, deep discharging, and thermal runaway—

conditions that can significantly degrade battery performance or lead to safety hazards. The BMS also balances the charge across individual cells to maintain uniformity and prolong overall battery life. Additionally, it communicates with other vehicle systems such as the powertrain controller and charger interface, enabling dynamic energy management during acceleration, regenerative braking, and charging events. In essence, the BMS acts as the brain of the battery system, safeguarding its health, optimizing performance, and ensuring longevity under varying operating conditions.

4.2 Battery-Based Energy Storage System

Battery-based energy storage is the most widely adopted solution for electric and hybrid vehicles due to its versatility, energy efficiency, and scalability. The dominant technology in this category is the Lithium-Ion (Li-ion) battery, which offers a high energy density, long cycle life, and strong performance characteristics. Li-ion batteries have become the standard in modern electric vehicles due to their compact size, relatively light weight, and efficient energy conversion, typically achieving round-trip efficiencies of 90–95%. In addition to Li-ion, Nickel-Metal Hydride (NiMH) batteries are used in some hybrid vehicles, offering good durability and reliability, albeit with slightly lower energy density. Older technologies such as Lead-Acid batteries, though once prevalent, are now largely obsolete in traction applications due to their low energy density and heavy weight. Despite their advantages, battery systems are not without limitations. A major concern is capacity degradation over time, especially with repeated charge-discharge cycles, which leads to reduced driving range and performance. Furthermore, thermal runaway—a condition where increased temperature leads to uncontrollable heat generation—poses significant safety risks in Li-ion systems if not properly managed. Therefore, advanced thermal management and battery management systems (BMS) are critical to ensure safe and efficient operation. Overall, battery-based energy storage remains a cornerstone of modern EV and HEV design, balancing performance, safety, and cost considerations.

Principle of Lead-acid battery

The working of a lead-acid battery is based on electrochemical reactions between lead (Pb), lead dioxide (PbO_2), and sulphuric acid (H_2SO_4).

During Discharge (Supplying Power)

Positive Plate Reaction:

$$PbSO_4(s) + 2H_2O(l) \rightarrow PbO_2(s) + SO_4^{2-}(aq) + 4H^+(aq) + 2e^-$$

Negative Plate Reaction:

$$PbSO_4(s) + 2e^- \rightarrow Pb(s) + SO_4^{2-}(aq)$$

Overall cell reaction (Charge)

The overall reaction for the charge cycle is the reverse of the discharge reaction.

$$2PbSO_4(s) + 2H_2O(l) \rightarrow Pb(s) + PbO_2(s) + 2H_2SO_4(aq)$$

Sulphuric acid concentration decreases, and water content increases.

During Charging (Restoring Energy)

- Current is passed in the reverse direction.
- Lead sulphate ($PbSO_4$) on plates converts back to **PbO_2** (positive) and **Pb** (negative).
- Electrolyte regains its original concentration

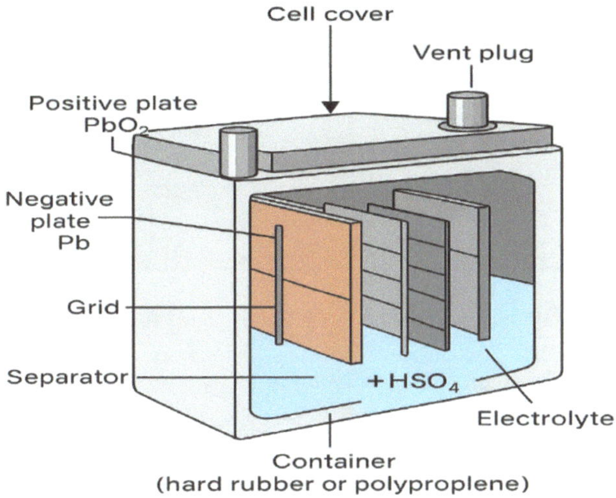

Fig.4.4: Construction & working of lead-acid battery

4.3 Analysis Factors GTUexam-3170923-WIN2024

Key Factors Affecting Battery Performance and Longevity are the performance, efficiency, and lifespan of battery-based energy storage systems in electric and hybrid vehicles are significantly influenced by several interrelated factors. One of the most critical is the State of Charge (SOC), which indicates the current level of energy stored in the battery relative to its full capacity. Closely related is the Depth of Discharge (DOD), representing how much of the battery's energy has been used. Operating batteries within an optimal SOC window and limiting deep discharges (high DOD) can substantially extend battery life and improve thermal management, as battery performance is highly sensitive to temperature. Excessive heat can accelerate chemical degradation and increase the risk of thermal runaway, while low temperatures can reduce power output and charging efficiency. To mitigate these effects, modern EVs employ active cooling or heating systems to maintain the battery within its ideal operating temperature range. Additionally, the charge/

Table 4.2: classification of batteries

Battery Type	Energy Density	Power Density	Cycle Life	Cost	Thermal Stability	Applications
Lithium-Ion (Li-ion)	High (150–250 Wh/kg)	High	1,000–2,000 cycles	Moderate–High	Medium (requires BMS)	Most EVs (e.g., Tesla, Nissan)
Nickel-Metal Hydride (NiMH)	Medium (60–120 Wh/kg)	Moderate	300–500 cycles	Moderate	High	HEVs (e.g., Toyota Prius)
Lead-Acid	Low (30–50 Wh/kg)	Low–Moderate	200–300 cycles	Low	Low	Rare, legacy or auxiliary use
Solid-State (emerging)	Very High (>300 Wh/kg)	High	Potentially >2,000	Very High (currently)	Very High (stable)	Future EVs (R&D/ Prototype)

discharge rate, commonly referred to as the C-rate, plays a pivotal role in determining battery health and performance. High C-rates—indicating fast charging or discharging—can cause internal heating and accelerate wear, whereas moderate C-rates promote longer cycle life and stable operation. Therefore, maintaining balanced operating conditions in terms of SOC, temperature, and C-rate is essential for ensuring safe, reliable, and long-term battery operation in electric mobility systems.efficiency. Another crucial factor is

- State of Charge (SOC) & Depth of Discharge (DOD)
- Thermal management
- Charge/discharge rate (C-rate)

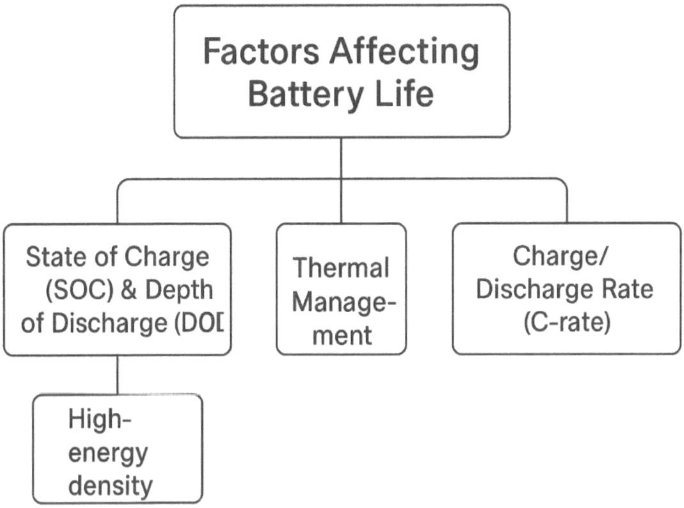

Fig.4.5: Battery Life

4.4 Typical Specifications of EV Batteries

Modern electric vehicle (EV) battery systems are designed to balance performance, longevity, and safety within practical constraints of cost and weight. A key performance metric is energy density, which typically ranges between 150 and 250 Wh/kg for lithium-ion batteries used in contemporary EVs. This allows for sufficient energy storage to support driving ranges that meet consumer expectations while maintaining manageable battery pack sizes and weights. Another crucial parameter is cycle life, which refers to the number of complete charge-discharge cycles a battery can undergo before its capacity significantly degrades. For high-quality lithium-ion chemistries, the typical cycle life ranges from 1,000 to over 2,000 cycles, depending on usage patterns, thermal conditions, and depth of discharge. These specifications are central to determining the overall performance and economic viability of EVs, as they directly influence driving range, maintenance schedules, and total cost of ownership.

- Energy density: 150–250 Wh/kg
- Cycle life: 1000–2000

EV Battery Specifications Comparison Table

Table 4.3: Specifies EV battery comparison on the base of their capacity

EV Model / Battery Type	**Energy Density** *(Wh/kg)*	**Cycle Life** *(cycles)*	**Battery Capacity** *(kWh)*
Tesla Model 3 (Li-ion NCA)	250	1500	60.0
Nissan Leaf (Li-ion LMO/NMC)	150	1000	40.0
Hyundai Kona EV (Li-ion NCM)	180	1200	64.0
BYD Han EV ($LiFePO_4$)	160	3000	76.9
Lucid Air (Li-ion NMC)	270	2000	113.0
Chevy Bolt (Li-ion NMC)	160	1500	66.0

4.5 Fuel Cell-Based Energy Storage Technology

4.5.1 Analysis Factors in Fuel Cell Energy Systems

Fuel cell systems, particularly Proton Exchange Membrane Fuel Cells (PEMFCs) used in electric and hybrid vehicles, involve several critical design and integration factors that influence their performance, efficiency, and suitability for automotive applications. One of the primary technical considerations is the stack voltage-current characteristic, which defines the relationship between the operating voltage and the current output of the fuel cell stack. As current increases, voltage typically drops due to activation, ohmic, and concentration losses—affecting both power output and system efficiency. Accurate modelling and control of these characteristics are essential to ensure optimal operation under varying load conditions.

Another key aspect is the design of the hydrogen storage tank, which must safely contain high-pressure hydrogen (typically 350–700 bar) while minimizing weight and maximizing volumetric efficiency. Advanced composite materials and multi-layer designs are employed to meet safety standards without compromising storage capacity. The overall vehicle layout must accommodate these tanks without affecting space or crash safety. Equally important is the integration of the fuel cell system with a secondary energy storage unit, usually a lithium-ion battery. This hybridization enables load-levelling, regenerative braking, and power buffering during transients such as acceleration or hill climbs. The battery absorbs energy during regenerative events and supplements the fuel cell during high-demand conditions, enhancing overall system responsiveness. In terms of performance metrics, fuel cell systems offer efficiencies between 40% to 60%, depending on load and operating conditions. However, they generally suffer from lower power density, typically around 1 KW/L, which poses challenges in compact vehicle design. Despite these limitations, the combination of fuel cell stacks with batteries provides a promising pathway for zero-emission, long-range electric mobility, especially in heavy-duty and long-haul applications. Fuel cell–battery hybrid systems are particularly suitable for long-haul transportation, where extended driving range and rapid refueling are critical. Long-haul" refers to transportation over long distances, typically involving extended travel time and high energy demand.

4.5.2 Fuel Consumption Analysis as per research in the University of Indonesia

Energy Consumption Analysis of Hybrid Electric Vehicles in Urban Tropical Conditions

Research conducted at the University of Indonesia identifies the Internal Combustion Engine (ICE) as one of the largest consumers of fossil fuels. Although extensive efforts have been made to replace ICE-based propulsion systems with electric motors and battery-based energy storage, complete elimination of ICE technology remains impractical in the near term. A viable intermediate solution is the integration of an ICE with an electric motor, resulting in a Hybrid Electric Vehicle (HEV). This study aims to quantify the energy consumption of HEVs operating in urban environments within a tropical megapolitan city. The methodology involves direct monitoring of energy usage in hybrid vehicles under real-world driving conditions. Sensors and data acquisition systems are installed on the test vehicles to record key operational parameters, including route characteristics, vehicle speed, fuel consumption, and electrical energy usage. For benchmarking purposes, similar data acquisition systems are also installed on conventional ICE vehicles with comparable engine capacities. In addition, findings from previous studies on Plug-In Hybrid Electric Vehicles (PHEVs) are incorporated for comparative analysis.

The results demonstrate that HEVs are capable of traveling approximately 22 km per liter of gasoline, whereas conventional ICE vehicles achieve only about 13 km per litre. In comparison, PHEVs exhibit the highest efficiency, covering up to 34 km per litre of gasoline. These findings indicate that HEVs offer significant improvements in fuel efficiency over conventional ICE vehicles, although PHEVs provide the greatest reduction in gasoline consumption among the three configurations. The primary focus of this research is an evaluation of fuel consumption in HEVs. Furthermore, the data collected through the onboard data logger will be utilized for subsequent analyses and extended studies related to vehicle energy performance

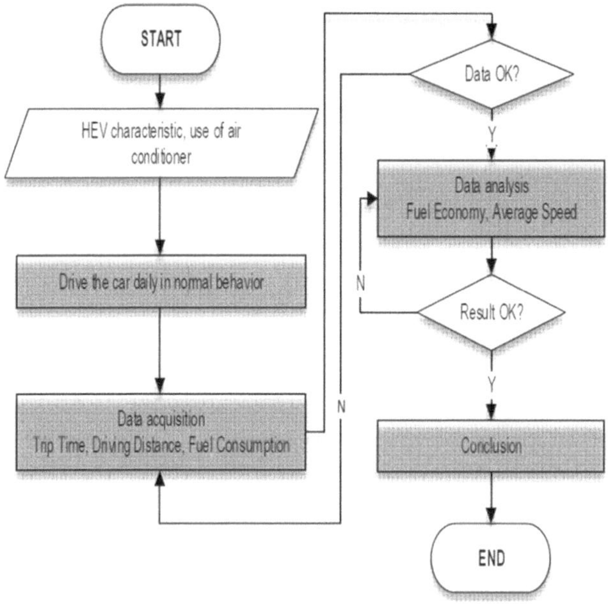

Fig.4.6: Experimental flow charts

Range	Trip Time [min]	Driving Distance [km]	Fuel Consumption [L]	Fuel Economy [km/L]	Average Speed [km/h]
0-10 km	266.8	32.1	3.7	8.6	7.23
10-20 km	634.4	132.8	7.8	17.1	12.56
20-30 km	519.4	102.7	6.0	17.0	11.87
30-40 km	1119.5	281.0	14.3	19.7	15.06
40-50 km	302.7	92.2	4.4	20.8	18.28
50-60 km	624.0	224.4	10.0	22.3	21.58
60-70 km	445.5	138.3	6.1	22.6	18.63
70-80 km	3073.5	1266.6	54.8	23.1	24.73
80-90 km	1897.2	689.7	30.7	22.4	21.81
90-100 km	287.6	97.2	4.4	22.3	20.28
100-110 km	571.6	214.4	8.6	24.9	22.51
>110 km	5087.2	2356.9	105.4	22.4	27.80
Overall	14829.2	5628.3	256.4	21.95	22.8

Fig.4.7: HEV energy consumption – experiment data of Prius

Analysis of Fuel Consumption over Driving Range

Collective information was gathered over three months, from 26th June to 28th September. As summarized in Table, fuel consumption varied significantly across different driving ranges. In the 0–10 km range, fuel efficiency was the poorest, with a consumption rate of 8.6 km/L. The cumulative distance travelled within the 0–30 km range exceeded 200 km, with the highest fuel efficiency in this range recorded at 17.1 km/L. Additionally, more than 1200 minutes of travel time were required to cover distances within the 0–30 km range. For the 30–60 km range, the total distance travelled was approximately 600 km, with a cumulative travel time of about 2000 minutes. The most efficient fuel consumption was observed in the 60–90 km range, achieving 23.1 km/L. In the 90–110 km range, the total mileage was less than 320 km, with peak fuel efficiency occurring between 100 and 110 km at 24.9 km/L. The highest average vehicle speed, 27.8 km/h, was recorded for distances exceeding 110 km. In this range, the average fuel consumption was 22.4 km/L, with a total cruising distance exceeding 2300 km. Overall, the data indicate that lower-speed, short-distance operation (0–10 km) results in substantially higher fuel consumption, whereas optimal fuel efficiency is achieved in the 100–110 km range.

Across the entire experimental period, the vehicle accumulated a total mileage exceeding 5600 km, with an average fuel consumption of 21.95 km/L. Analysis of driving behavior further indicates that the longest cumulative travel distances occurred within the 70–80 km and 80–90 km ranges.110 km is ignored; then the longest distance was in the range of 70-80 km. It was 1266.6 km. The results of experiments in dense tropical regions illustrate that actual fuel consumption is higher than the claims of the brochure. The air conditioning is mandatory in the tropics. It should be suspected that energy is also channeled large enough to cool the cabin. It is consistent with previous research, which states that AC is the second largest power consumption after the main motor in tropical electric vehicles. The use of HEV in big city areas with almost always traffic jams, like Jakarta, only gives a small added value to fuel savings. The dominant driving distance of the driver in this experiment is between 70-80 km, and in this range also shows the most efficient fuel consumption. The range can be a reflection of the distance from home- office and offices every day and represents the distance from Bekasi - Jakarta, Bogor-Jakarta, and Tangerang to Jakarta. For tropical areas and big cities like Jakarta, traffic conditions are not favourable. But if HEV is used to cover medium distances (70-80 km), the benefits will be optimal. In subsequent studies, it is very strategic to discuss the relationship between HEV, PHEV, BEV prices, local fuel prices (Indonesia), taxation, and battery prices. The research is essential so that it can provide suggestions to policyholders, consumers, and producers related to the type and price of the best electric vehicles for developing countries. Greenhouse issues, scarcity and rising fuel prices, electricity unit prices, and the amount of tax will determine the acceptance of electric cars in Indonesia

Analysis of Proton Exchange Membrane Fuel Cells – Widely used in automotive

Pros: Very high energy density (hydrogen), Only emission = water, Refill like gasoline

Cons: Expensive (platinum catalyst), Storage of H_2 is tricky (700 bar tanks!)

Analysis Factors: Stack voltage-current characteristics, Hydrogen tank design, Integration with battery

Efficiency: 40–60%, Power density: ~1 kW/L

Supercapacitor-Based Energy Storage, Also Called: Ultracapacitors

Pros: Ultra-fast charge/discharge, long life (>1 million cycles), Excellent for regenerative braking

Cons: Low energy density (~5–10 Wh/kg), Self-discharge is high

Analysis Factors: ESR (Equivalent Series Resistance), Charge/discharge curves, Series-parallel design for voltage/current match It is used for: *Power assist, hybridization*

4.6 Flywheel-Based Energy Storage

Flywheel-based energy storage systems operate by converting and storing energy in the form of rotational mechanical energy. The fundamental concept revolves around a rotor or flywheel that spins at very high speeds—sometimes up to tens of thousands of revolutions per minute (RPM). When electrical energy is available, it is used to accelerate the rotor using an electric motor; during discharge, the spinning rotor drives a generator to produce electricity. This mechanism enables extremely fast charge and discharge cycles, making flywheels ideal for applications requiring quick bursts of high power. A key strength of flywheels lies in their high power density and long operational life, as they are resistant to issues like deep discharge, which typically degrade chemical batteries. Additionally, they offer impressive efficiency (up to 95%) and can undergo thousands of cycles without significant performance loss. Their robustness and rapid response time make them attractive for energy recovery systems, especially in high-performance vehicles like Formula 1 cars and hybrid buses. However, these advantages come with certain trade-offs. Flywheels tend to be heavy, primarily due to the mass required to store sufficient rotational energy. Moreover, the high-speed rotation introduces challenges such as the gyroscopic effect, which can influence vehicle dynamics and stability. To minimize friction and aerodynamic drag, flywheels often operate in a vacuum chamber with magnetic or air bearings, which adds complexity and cost. From a design perspective, safety is a major consideration, as failure at high rotational speeds can be catastrophic—thus requiring robust enclosures and careful control of rotor inertia and maximum RPM. In terms of energy storage capacity, flywheels offer moderate energy density, typically in the range of 20–80 Wh/kg, which is lower than batteries but significantly higher than supercapacitors. They are not suitable for long-term energy storage but excel in short-duration, high-power applications. As such, flywheels are often used in combination with other energy storage technologies to form hybrid systems, where they manage peak loads and support rapid energy transfer needs.

Fuel Cell and Flywheel as Energy Source Elements in Electric and Hybrid Electric Vehicles GTUexam-3170923-WIN2024

In electric and hybrid electric vehicles (EVs and HEVs), fuel cells and flywheels serve as alternative or supplementary energy sources, improving performance, efficiency, and energy recovery. A fuel cell is an electrochemical device that converts the chemical energy of hydrogen and oxygen directly into electrical energy, with water and heat as by-products. A fuel cell primarily comprises an anode, a cathode, and an electrolyte. At the anode, hydrogen molecules undergo electrochemical dissociation into protons and electrons. The protons migrate through the electrolyte, while the electrons are directed through an external circuit, thereby generating electrical power. At the cathode, oxygen combines with the arriving protons and electrons to produce water as the electrochemical reaction. Fuel cells offer high energy conversion efficiency, emit no harmful pollutants—producing only water vapor—and are capable of delivering continuous electrical power provided that the fuel supply is maintained. In EVs and HEVs, fuel cells are used either as a primary energy source, as in Fuel Cell Electric Vehicles (FCEVs), or in combination with batteries or supercapacitors to enhance efficiency and range. A flywheel is a mechanical energy storage system that stores energy in the form of rotational kinetic energy. It comprises a high-speed rotor, a motor-generator unit, and low-friction bearings often housed in a vacuum chamber to minimize energy losses. During braking or low power demand, the motor functions as a generator to store energy in the flywheel by spinning it at high speed. During acceleration or peak load, the stored kinetic energy is converted back into electrical energy to support propulsion. Flywheels offer high power density, rapid charge and discharge capability, and excellent regenerative braking performance.

They are particularly useful for handling short-term, high-power demands and reducing stress on batteries by assisting during peak load conditions.

4.7 Why Hybridize Energy Storage Systems?

In electric and hybrid electric vehicles, no single energy storage technology can independently fulfil all performance requirements—such as long driving range, rapid acceleration, regenerative braking, and sustained power delivery. Each storage device comes with its own strengths and limitations in terms of energy density, power density, charge/discharge rate, and cycle life. As a result, hybridizing multiple energy storage systems has emerged as a practical and efficient solution to overcome these individual limitations and to optimize vehicle performance across diverse driving conditions. For instance, combining a battery with a supercapacitor enables the system to benefit from the high energy density of the battery for extended range, while the high-power density and rapid response of the supercapacitor support quick acceleration and efficient regenerative braking. Similarly, pairing a fuel cell with a battery merges the long-term sustainability and continuous energy generation of hydrogen fuel cells with the dynamic responsiveness and energy buffering capabilities of batteries, making it suitable for both steady-state cruising and peak power demands. The effectiveness of any hybrid energy storage system, however, depends heavily on the control strategy employed. Intelligent control algorithms are essential to manage the power flow between different storage devices in real time, ensuring that each component operates within its optimal performance window. These strategies determine when to draw energy from each source, when to recharge them, and how to handle rapid transitions in load demand, thereby maximizing system efficiency, safety, and longevity.

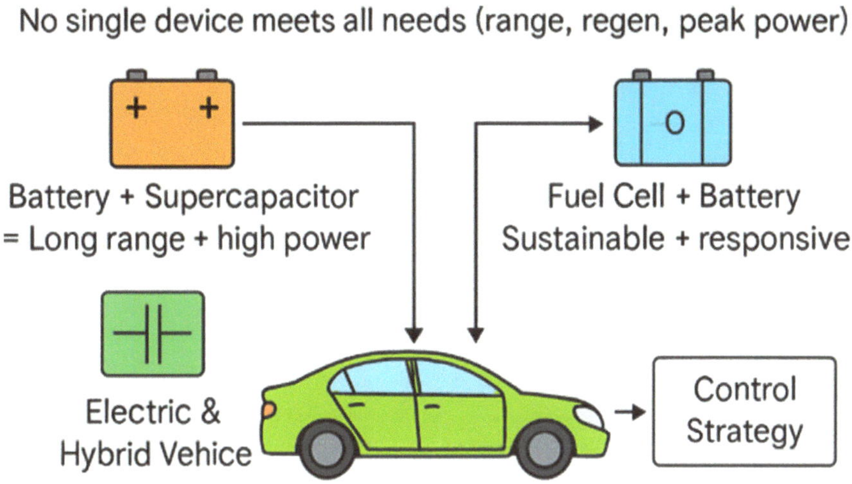

Fig.4.8: Why hybridize?

4.8 Sizing the Drive System in EV/HEV

Proper sizing of the drive system in electric (EV) and hybrid electric vehicles (HEV) is critical to ensuring optimal performance, energy efficiency, and system reliability. The drive system typically consists of the propulsion motor, power electronics, and in the case of HEVs, an internal combustion engine (ICE) that works in conjunction with the electric motor. The design must cater to a variety of operating conditions such as cruising, acceleration, climbing gradients, and regenerative braking.

Matching the Electric Machine and ICE (for HEVs)

In hybrid configurations, both the ICE and electric motor must be carefully coordinated to deliver the required torque and power across the speed range. This is achieved by overlapping their torque-speed envelopes to ensure they complement each other during various drive cycles. Additionally, the load demand profile of the vehicle—derived from its intended application (urban, highway, off-road, etc.)—guides how each component contributes during start-up, cruising, and transient events.

Numerical Example:

- Vehicle mass $m = 1200\,\mathrm{kg}$
- Top speed $V = 120\,\mathrm{km/h} = 33.3\,\mathrm{m/s}$
- Gradeability = 10% ($\theta \approx 5.71^\circ$)
- Acceleration: $0 \rightarrow 50\,\mathrm{km/h}$ in 8 seconds → $a = \frac{13.89}{8} \approx 1.74\,\mathrm{m/s}^2$
- Rolling resistance $F_r \approx 0.015 \cdot mg$
- Efficiency $\eta = 0.9$

$$F = m \cdot a + mg \cdot \sin(\theta) + F_r$$

$$= 1200 \cdot 1.74 + 1200 \cdot 9.81 \cdot \sin(5.71^\circ) + 0.015 \cdot 1200 \cdot 9.81$$

$$= 2088 + 1170 + 177 \approx 3435\,\mathrm{N}$$

Now calculate

$$P = \frac{F \cdot v}{\eta} = \frac{3435 \cdot 33.3}{0.9} \approx 127,100\,\mathrm{W} \approx 127\,\mathrm{kW}$$

Result: ~40–60 KW motor required for urban EV. For conservative, urban use (lower speeds, partial acceleration), typical motor sizes range from 40–60 KW.

4.9 Sizing Power Electronics

After determining the specifications of the propulsion motor, the next crucial step is to appropriately size the power electronic components, ensuring they can manage the system's electrical demands with high efficiency, reliability, and safety. The inverter, which converts DC from the battery to AC for the motor, must be rated to match or slightly exceed the motor's voltage and current requirements. To accommodate electrical transients—such as sudden acceleration, deceleration, or switching surges—a design margin of 20–30% is typically incorporated into the inverter's power handling capacity.

In addition to the inverter, other essential components like DC-DC converters are selected based on the system's power stage efficiency, thermal constraints, and the specific voltage levels required by auxiliary systems (e.g., lighting, infotainment, control units). Effective thermal management is imperative, as power electronics generate significant heat during operation. Components must also adhere to design standards concerning compactness, electromagnetic compatibility (EMC), and system-level safety protocols. The overall goal in power electronics sizing is to ensure robust performance under variable loads while minimizing energy losses and maintaining component longevity—particularly in the dynamic and often harsh environments experienced by electric and hybrid vehicles.

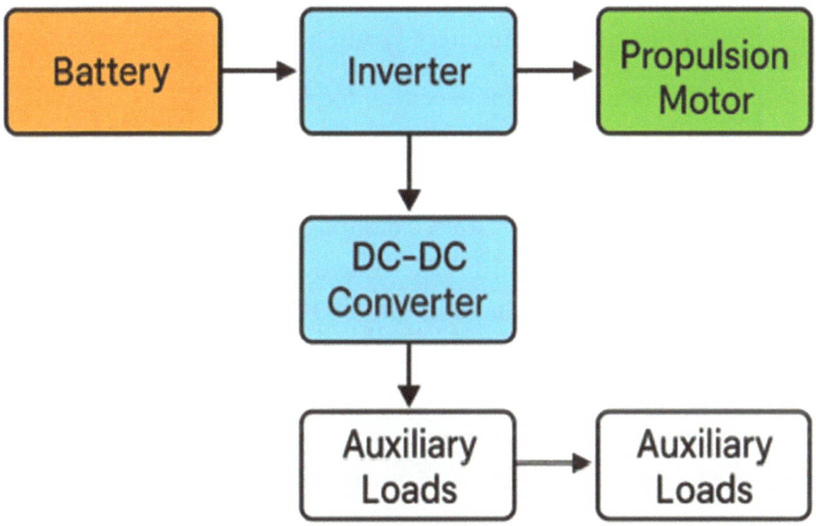

Fig.4.9: Sizing the power electronics

The battery is the primary energy storage device in electric and hybrid vehicles. It stores DC electrical energy and supplies it to the drive system. Typical batteries in EVs are Lithium-ion due to their high energy density, cycle life, and efficiency. The battery must be sized based on vehicle energy demand, voltage level, and thermal constraints. The inverter converts the DC power from the battery into AC power needed to drive the AC propulsion motor (commonly PMSM or induction motor). It also performs regenerative braking by converting AC power generated by the motor back to DC to recharge the battery. The inverter plays a key role in controlling motor speed and torque through techniques like PWM (Pulse Width Modulation). Must match the motor's voltage/current ratings, must handle transients, hence 20–30% headroom is added, Efficiency and thermal management are key design considerations. The propulsion motor converts the electrical energy into mechanical energy to drive the wheels. This is typically a high-efficiency AC motor (e.g., Permanent Magnet Synchronous Motor or Induction Motor). It provides the required torque and speed under various driving conditions such as acceleration, climbing, or cruising. DC to DC converter steps down the high-voltage DC from the main battery (e.g., 400–800 V) to a lower voltage level (e.g., 12 V or 48 V) for auxiliary components. It ensures safe and efficient operation of Auxiliary Loads.

These are the non-propulsion electrical loads in the vehicle, powered by the low-voltage bus. While not directly involved in vehicle movement, they are essential for safety, user comfort, and vehicle control. Examples include: Headlights, indicators, dashboard, Steering and braking assistance, cooling fans, and HVAC systems. These components require a reliable and regulated voltage from the DC-DC converter to function correctly. Common applications are Lighting systems, Control units (ECU, BMS, sensors), Air conditioning, infotainment, wipers, etc. It often includes isolation, protection circuitry, and power stage efficiency optimization.

Table for selecting Energy Storage Tech

Table 4.4: Energy storage technology

Criteria	**Battery**	**Fuel Cell**	**Supercap**	**Flywheel**
Energy density	High	Very high	Low	Medium
Power density	Medium	Low-Medium	High	High
Response time	Secs	Secs	ms	ms
Cost	Medium	High	Medium	High
Cycle life	1,000–3,000	5,000	1M+	10,000+

Calculating Ratings

To accurately assess the performance requirements and energy demands of an electric vehicle (EV) or hybrid system, standardized drive cycles such as the New European Driving Cycle (NEDC) or the Worldwide Harmonized Light Vehicles Test Procedure (WLTP) are employed. These drive cycles simulate real-world driving conditions by specifying time-velocity profiles over a defined period. By analysing vehicle behaviour over a given cycle, engineers can compute several key performance metrics. Peak power is determined by identifying the maximum instantaneous power demand during aggressive acceleration or high-speed operation segments. In contrast, average power is calculated by integrating the power over the entire cycle and dividing it by the total cycle time, offering insight into typical system loading. Additionally, the total energy required per cycle is obtained by integrating the power consumption over the full duration of the drive cycle, which directly influences battery sizing and energy management strategies. Finally, the thermal load profile is derived from time-varying power loss data (e.g., inverter/motor losses), revealing how much heat is generated and thus informing the design of the thermal management system.

Power Electronics in HEV -Use of chopper-controlled DC motor drives *GTUexam-3170923-WIN2024*

If we understand the two-quadrant operation of the Chopper DC Motor Drive, we conclude that In electric vehicle (EV) applications, chopper-controlled DC motor drives are used for efficient speed control and bidirectional power flow. The two-quadrant operation allows the motor to operate in motoring mode (forward power flow) and regenerative braking mode (reverse power flow), while maintaining unidirectional current flow through the motor.

In Quadrant I, known as the forward motoring mode, the chopper regulates the average voltage applied to the motor armature by adjusting its duty cycle. In this mode, both torque and current remain positive, and power flows from the source to the motor. As a result, the motor provides the necessary torque to accelerate the vehicle in the forward direction, making this the primary mode for propulsion during normal driving conditions.

In Quadrant II, known as the regenerative braking mode, the motor functions as a generator during braking, converting the vehicle's mechanical energy back into electrical energy. In this mode, power flows from the motor to the battery or power source, enabling energy recovery and improving overall efficiency. The torque remains positive while the back electromotive force (EMF) exceeds the supply voltage, ensuring controlled and effective braking performance.

EV Energy Storage Overview

Table 4.4: Comparison of storage type

Storage Type	Energy Density (Wh/kg)	Power Capability (W/kg)	Typical Response Time	Cycle Life
Lithium-Ion Battery	150–250	1,000–3,000	~1 s	1,000–3,000 cycles
Hydrogen Fuel Cell	800–1,000 (H_2)	500–1,000	~2–5 s	4,000–6,000 cycles
Supercapacitor	5–10	10,000+	Milliseconds	1,000,000+ cycles
Flywheel	20–80	1,000–5,000	Milliseconds	10,000+ cycles

Multiple Choice Questions

Q1. What characteristic must an EV energy storage system prioritize to achieve an extended driving range?

A. High power density

B. Low voltage operation

C. High energy density ☑

D. High internal resistance

Why: Greater energy per kilogram means more stored charge, enabling longer trips.

Q2. Ensuring compatibility with regenerative braking allows the system to:

A. Increase the vehicle's top speed

B. Enable quicker AC charging

C. Capture kinetic energy during deceleration ☑

D. Shrink the traction motor

Why: Regen braking converts motion energy into electrical energy and stores it back.

Q3. The Battery Management System (BMS) protects Li-ion cells from which hazardous scenario?

A. Capacitive discharge

B. Thermal runaway ☑

C. Ohmic heating

D. Supercooling

Why: BMS controls temperature, voltage, and current to prevent runaway reactions.

Q4. Which parameter is not typically tracked by a BMS?

A. State of Charge (SOC)

B. Inverter frequency ☑

C. Cell temperature

D. Current flow

Q5. Which battery chemistry dominates today's EV market (e.g., Tesla, Nissan Leaf)?

A. NiMH

B. Lead-Acid

C. Solid-State

D. Lithium-Ion ☑

Q6. What's the biggest limitation of Lead-Acid batteries in EVs?

A. Low cost

B. High mass and poor energy density ☑

C. Strong thermal behavior

D. Long cycle life

Q7. Which battery type is projected to lead future EV designs due to high energy density and stability?

A. Lead-Acid

B. NiMH

C. Solid-State ☑

D. LFP

Q8. Operating a battery in moderate SOC ranges and with low depth-of-discharge increases:

A. Motor speed

B. Battery cycle life ☑

C. Vehicle acceleration

D. Regeneration frequency

Q9. Why do EVs employ thermal management systems for their batteries?

A. To cut manufacturing costs

B. To upgrade infotainment performance

C. To maintain safe cell temperatures ☑

D. To improve regen braking

Q10. Which EV model (from the given comparison) uses the highest energy-density battery?

A. BYD Han EV
B. Tesla Model 3 ☑
C. Nissan Leaf
D. Chevy Bolt

Q11. Typical energy density range for Li-ion EV packs is:

A. 30–50 Wh/kg
B. 60–120 Wh/kg
C. 150–250 Wh/kg ☑
D. 250–400 Wh/kg

Q12. Hydrogen fuel cells emit only:

A. CO_2
B. NOx
C. Water (H_2O) ☑
D. Methane

Q13. A common limitation of fuel cells is their:

A. Low energy density
B. Low power density ☑
C. High self-discharge
D. Strong high-C-rate ability

Q14. In a fuel cell hybrid powertrain, the battery primarily:

A. Stores hydrogen
B. Replaces the fuel cell
C. Handles regen energy and peak power bursts ☑
D. Cools the fuel stack

Q15. Flywheels usually fall within which energy density band?

A. 5–10 Wh/kg
B. 20–80 Wh/kg ☑
C. 150–250 Wh/kg
D. 800–1000 Wh/kg

Q16. A major engineering challenge for flywheels is:

A. Thermal control

B. High-voltage hazard

C. Mechanical containment due to gyroscopic forces ☑

D. Internal resistance

Q17. Why do modern EVs increasingly use hybrid storage systems?

A. To reduce cost only

B. To limit regenerative braking

C. To combine the strengths of multiple storage devices ☑

D. To eliminate power electronics

Q18. Typical design margin added when sizing an EV inverter is:

A. 5–10%

B. 10–15%

C. 20–30% ☑

D. 40–50%

Q19. Which component converts the EV's high-voltage DC bus into 12V for accessories?

A. BMS

B. Inverter

C. DC-DC Converter ☑

D. Rectifier

Q20. The EV inverter's core job is to:

A. Control dashboard indicators

B. Turn AC into DC

C. Convert DC battery power into AC for the motor ☑

D. Manage battery temperatures

India-GTU Questions

Question	Reference	Marks
Define following terms (1) Specific energy (2) Specific power (3) Energy density (4) Ambient temperature (5) Life cycle of battery (6) Energy stored (7) Charge capacity	chapt4pg113 SUM2025	**07**
Explain the difference between ultracapacitor and battery as an energy storage device for EV.	chapt4pg118	**07**
Define the terms charge capacity, specific energy, energy density, specific power, charge efficiency, energy efficiency, Life cycle of battery.	chapt4	**07**
Explain the basic principle of Super Capacitors based Energy Storage System in Hybrid Electric Vehicles?	chapt4pg116	**07**
Explain the two-quadrant operation of chopper DC motor drive with suitable waveforms for electric vehicle.	chapt4pg137 WIN2024	**07**
With neat diagram explain the construction and working of lead-acid battery.	chapt42pg121	**07**
Explain the fuel cell and flywheel as energy source elements in electric and Hybrid electricvehicles.	chapt5 chapt4pg131,136 GTUexam-3170923-WIN2024	**07**

Chapter 05

Modelling of Hybrid Electric Vehicle (HEV) Range

5 MODELLING OF HYBRID ELECTRIC VEHICLE (HEV) RANGE

5.1 Introduction to Vehicle Range Modelling

Vehicle range refers to the maximum distance a vehicle can travel on a full tank of fuel, a fully charged battery, or a complete energy replenishment cycle. In Hybrid Electric Vehicles (HEVs), the achievable range is determined by multiple factors, including the type of powertrain configuration, the characteristics of the energy source, the applied driving cycle, vehicle dynamics, and environmental conditions. Accurate range prediction is vital for energy management, battery sizing, and the development of optimal power split strategies between the internal combustion engine and the electric motor. One of the most reliable methods for estimating HEV range is through the use of driving cycles, which are standardized sequences of vehicle speeds and accelerations over time designed to replicate real-world driving conditions. Driving cycles are critical for assessing performance under controlled and repeatable scenarios, enabling engineers to simulate operational patterns such as stop-and-go traffic, steady cruising, regenerative braking opportunities, and variable load demands. Common examples include the New European

Driving Cycle (NEDC), the Worldwide Harmonized Light Vehicles Test Procedure (WLTP), and the Federal Test Procedure (FTP-75), each tailored to specific regional traffic profiles. In HEV modelling, driving cycles form the foundation for analysing battery State of Charge (SOC) variations, determining the proportion of electric versus combustion engine operation, and estimating regenerative energy recovery. Incorporating driving cycle data into simulation platforms allows for comprehensive evaluation of vehicle configurations, ultimately guiding design optimisation for extended range, enhanced efficiency, and reduced environmental impact.

Vehicle range refers to the maximum distance a vehicle can travel on a full tank of fuel, a fully charged battery, or a complete energy replenishment cycle. In Hybrid Electric Vehicles (HEVs), the achievable range is determined by multiple factors, including the type of powertrain configuration, the characteristics of the energy source, the applied driving cycle, vehicle dynamics, and environmental conditions. Accurate range prediction is essential, as it supports effective energy management, enables optimal battery sizing, and informs power split strategies between the internal combustion engine and the electric motor. One of the most reliable methods for estimating HEV range is through the use of driving cycles, which are standardized sequences of vehicle speeds and accelerations over time designed to replicate real-world driving conditions. Driving cycles are critical for assessing performance under controlled and repeatable scenarios, enabling engineers to simulate operational patterns such as stop-and-go traffic, steady cruising, regenerative braking opportunities, and variable load demands. Common examples include the New European Driving Cycle (NEDC), the Worldwide Harmonized Light Vehicles Test Procedure (WLTP), and the Federal Test Procedure (FTP-75), each tailored to specific regional traffic profiles. In HEV modelling, driving cycles form the foundation for analysing battery State of Charge (SOC) variations, determining the proportion of electric versus combustion engine operation, and estimating regenerative energy recovery. Incorporating driving cycle data into simulation platforms allows for comprehensive evaluation of vehicle configurations, ultimately guiding design optimisation for extended range, enhanced efficiency, and reduced environmental impact.

5.2 Driving Cycles

Driving cycles are standardized time–speed profiles that represent real-world vehicle operating conditions. They are designed to replicate typical patterns of acceleration, cruising, deceleration, and idling encountered in urban, suburban, or highway environments. The primary purpose of driving cycles is to evaluate a vehicle's fuel economy, analyse exhaust emissions, and model its range under both standardized laboratory tests and actual road conditions. By providing a repeatable framework, driving cycles allow engineers to compare vehicle performance across different designs and technologies, assess compliance with regulatory standards, and optimise powertrain control strategies for improved efficiency and reduced environmental impact. Indian drive cycle has different modes of operation: Starting and acceleration mode, Constant speed or cruising mode, Deceleration or braking mode.

5.2.1 Staring and acceleration

Operating Modes of the Hybrid Energy Storage System

High Power Demand Condition

When the demanded power exceeds the maximum rated battery power (P_demand > P_batt(max)), the ultracapacitor (UC) supports the battery through the bidirectional DC/DC converter (BDC), provided that the ultracapacitor voltage and state of charge satisfy the constraints V_UC > V_UC(min) and SOC_UC > SOC_UC(min). Under these conditions, the BDC operates in boost mode, supplying an elevated voltage to the DC link to meet the load demand.

Constant Speed Operation

When the demanded power is less than or equal to the maximum battery power (P_demand ≤ P_batt(max)), the battery alone supplies the required power. In this mode, the BDC operates in buck mode only to charge the ultracapacitor when the UC state of charge falls below its minimum threshold. If the UC state of charge remains above this limit, the interface circuit remains inactive.

Deceleration and Braking Mode

During deceleration or braking, the regenerative power generated at the DC link is managed and distributed within the hybrid energy storage system (HESS) according to the state of charge of its components. The ultracapacitor is preferentially charged when V_UC < V_UC(max) and SOC_UC < SOC_UC(max). Battery charging is initiated only after the ultracapacitor reaches its maximum voltage and state-of-charge limits. This control strategy ensures efficient recovery of regenerative energy while maintaining safe operating conditions for both energy storage elements.

5.2.2 Differences between BEV, HEV, and FCEV GTUexam-3170923-SUM2025

BEVs are fully electric, HEVs use both a gasoline engine and an electric motor, and FCEVs use hydrogen fuel cells to create electricity for their electric motors. BEVs are charged by plugging in, HEVs self-charge their small batteries with a gasoline engine and regenerative braking, and FCEVs are refuelled with hydrogen.

Here's a breakdown of each vehicle type:

Battery Electric Vehicle (BEV)

A Battery Electric Vehicle (BEV) relies solely on a large battery pack as its power source to drive the electric motor. The battery is recharged by plugging the vehicle into an external power supply. Since BEVs do not use an internal combustion engine, they produce zero tailpipe emissions, making them environmentally friendly. Their main advantages include high energy efficiency and clean operation without local emissions. However, they require a well-developed charging infrastructure, and the recharging process generally takes longer compared to the quick refuelling times of fuel cell electric vehicles (FCEVs).

Hybrid Electric Vehicle (HEV)

A Hybrid Electric Vehicle (HEV) uses a combination of an internal combustion engine (ICE) and an electric motor, supported by a relatively small battery. Unlike plug-in hybrids, it does not need to be connected to an external power supply, as its battery is recharged through the ICE and regenerative braking. While HEVs are more efficient than conventional vehicles, they are not zero-emission like BEVs or FCEVs, since the ICE still produces exhaust gases. Their main advantages include good energy efficiency and an extended driving range due to the combined power system, but they have a limited electric-only range and continue to contribute to some level of emissions.

Fuel Cell Electric Vehicle (FCEV)

A Fuel Cell Electric Vehicle (FCEV) uses a fuel cell to generate electricity by combining hydrogen and oxygen, which then powers an electric motor. It is refueled by adding hydrogen to a storage tank, in a process very similar to filling up a conventional gasoline vehicle. Since the only by-product of the reaction is water vapor, FCEVs produce zero harmful tailpipe emissions. Their key advantages include fast refueling times that take just a few minutes, a long driving range, and clean operation without pollutants. However, they face challenges such as high production costs and the limited availability of hydrogen refueling infrastructure, which restricts their widespread adoption.

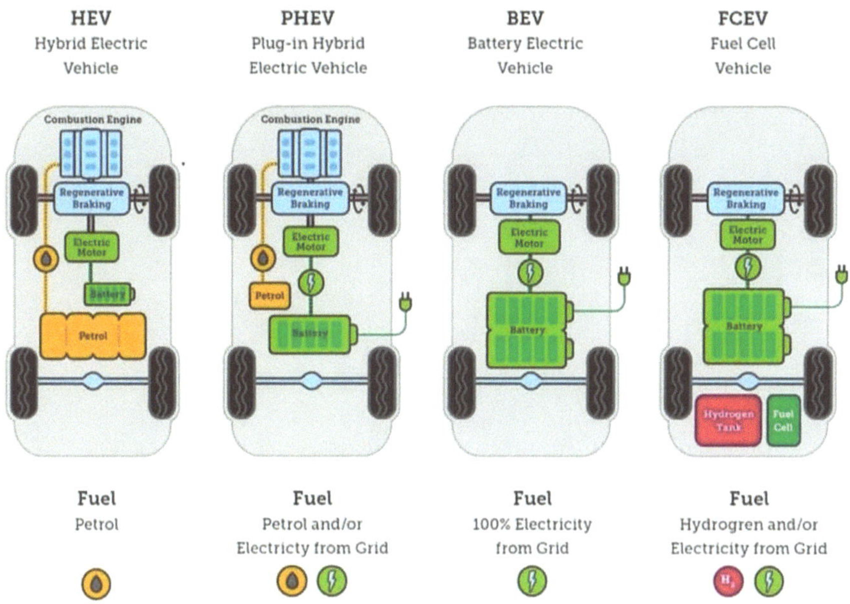

Fig.5.1: Difference between HEV, PHEV, BEV,FCEV

5.3 Types of Driving Cycles GTUexam-3170923-WIN2025

Driving cycles can be broadly classified into several categories based on their origin and application. Standard cycles are laboratory-based and designed to be reproducible, enabling consistent comparisons of vehicle performance; common examples include the Worldwide Harmonized Light Vehicles Test Procedure (WLTP), the New European Driving Cycle (NEDC), and the Federal Test Procedure (FTP-75). Real-world cycles are derived from on-road measurements, such as the Real Driving Emissions (RDE) test, and capture the variability of actual driving conditions. Custom or test-track cycles are developed for specific research or evaluation purposes, such as the Indian Driving Cycle (IDC), which reflects typical Indian traffic patterns. Application-specific cycles are tailored to particular vehicle classes or operational environments, including the World Motorcycle Test Cycle (WMTC) for two-wheelers and the Modified Indian Driving Cycle (MIDC) for light vehicles in India.

Each type serves a distinct role in evaluating vehicle efficiency, emissions, and range under targeted conditions, ensuring accurate modelling and performance analysis.

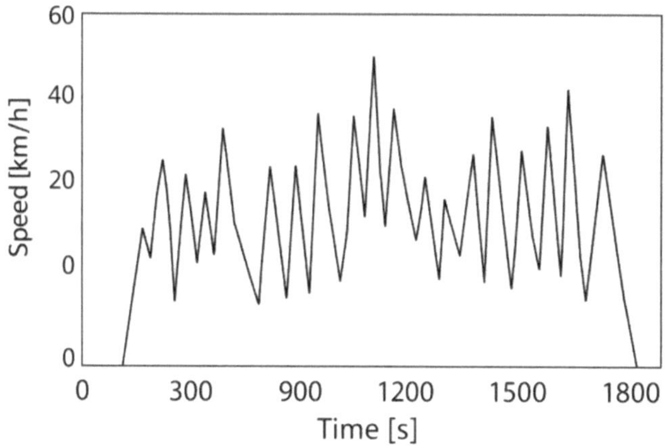

Fig.5.2: Speed vs Time for a typical urban driving cycle-FTP-75

Federal Test Procedure (FTP-75)

The Federal Test Procedure-75 (FTP-75) is a standardized driving cycle used in the United States for emission certification and fuel economy evaluation of light-duty vehicles. It is commonly referred to as the "FTP" and should not be confused with the FTP test procedures applied to heavy-duty engines.

FTP-75 is an extension of the Environmental Protection Agency (EPA) Urban Dynamometer Driving Schedule (UDDS) and is derived from the earlier FTP-72 cycle. The primary distinction between FTP-75 and FTP-72 is the inclusion of an additional third phase, which replicates the initial transient phase under hot-start conditions. This phase is conducted after a 10-minute engine shutdown, thereby accounting for emissions generated during hot restart operation.

The complete FTP-75 cycle consists of the following segments:

1. **Cold-start transient phase** (ambient temperature 20–30 °C): 0–505 s
2. **Stabilized phase**: 506–1372 s

3. **Hot soak period**: minimum 540 s, maximum 660 s

4. **Hot-start transient phase**: 0–505 s

Exhaust emissions from each phase are collected separately in Teflon sampling bags, analysed individually, and reported in grams per mile (g/mi) or grams per kilometer (g/km). The corresponding weighting factors applied to each phase are 0.43 for the cold-start transient phase, 1.0 for the stabilized phase, and 0.57 for the hot-start transient phase.

Key characteristics of the FTP-75 driving cycle are summarized as follows:

- **Total duration:** 1877 s
- **Total distance:** 11.04 miles (17.77 km)
- **Average speed:** 21.2 mph (34.12 km/h)
- **Maximum speed:** 56.7 mph (91.25 km/h)

For emission certification, vehicles are required to comply with the applicable FTP-75 emission standards. Beginning with the 2000 model year, vehicles must also undergo testing using two Supplemental Federal Test Procedures (SFTP) to address limitations of FTP-75. These include the US06 cycle, which represents aggressive high-speed driving, and the SC03 cycle, which evaluates emissions and fuel consumption associated with air-conditioning operation

CAFE fuel economy values are calculated on the basis of FTP and HWFET testing. Until model year 2007, EPA on-road fuel economy values shown on new vehicles' labels were calculated on the basis of FTP testing for the city rating, while the HWFET test was used for the highway rating. Since model year 2008, the FTP is used for the determination of the EPA on-road fuel economy ratings using the EPA 5-cycle method. The 5-cycle results are calculated based on the results of two FTP tests—one regular and one cold temperature test, run at a lab temperature of 20°F (-6.7°C)—as well as the HWFET, US06, and SC03.

The FTP-75 cycle is known in Australia as the ADR 37 (Australian Design Rules) cycle and in Brazil as test standard NBR6601.

A variant of the FTP-75 cycle—where the stabilized phase is run again after the completion of the hot start phase—is sometimes used in certain applications, for instance, in some hybrid vehicle tests.

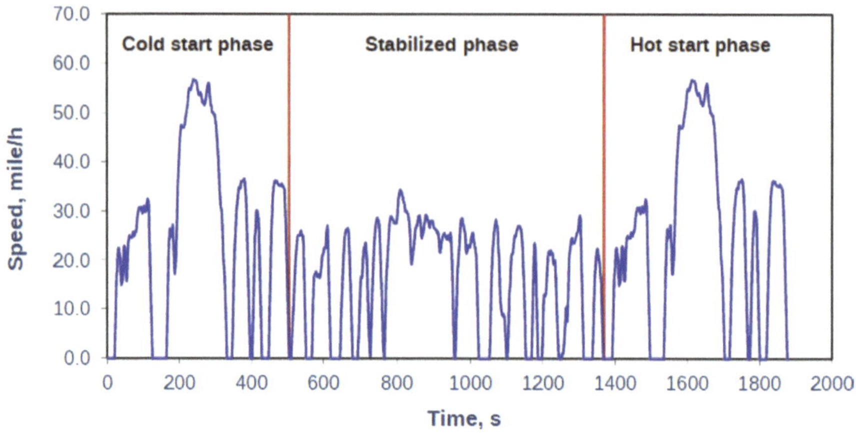

Fig.5.3: US EPA Urban Dynamometer Driving Schedule (FTP-75)

5.4 Range Modelling for Battery Electric Vehicles (BEVs)

The energy storage system (ESS) is a fundamental component of pure electric vehicles (EVs), with batteries serving as the most commonly adopted ESS. Despite their widespread use, battery energy storage systems (BESS) in pure EVs face several limitations, including restricted driving range, degradation of battery lifespan, and relatively low power density. These challenges have motivated researchers to explore the integration of multiple energy storage technologies to enhance overall vehicle performance. In addition to batteries, ultracapacitors (UCs) and fuel cells have emerged as promising energy storage alternatives for EV applications. A vehicle propulsion architecture utilizing a proton-exchange membrane fuel cell (PEMFC) as the primary energy source, supported by a nickel–metal hydride (Ni–MH) battery as a secondary source, has been proposed. However, fuel cell–based EV systems encounter significant technological constraints related to system weight, volumetric size, and cost, which must be mitigated to achieve desirable efficiency levels.

To further improve the operational performance of fuel cell–powered EVs, the incorporation of an auxiliary energy source based on ultracapacitors has been recommended. Reported studies demonstrate that such hybrid configurations offer notable advantages, particularly in terms of power density enhancement and dynamic performance, thereby validating the benefits of energy storage system hybridization.

S. no.	Year	Energy source	Description
1.	2005	Battery and ICE	This hybrid power train is proposed as an alternative to the ICE vehicle. The results of power assisting and range extension modes are found to be satisfactory.[9]
2.	2009	Battery and ICE	Developed micro-hybridized Bajaj RE2 indicates 21% reduction of consumption and emissions with improved driver comfort.[10]
3.	2010	Solar and battery	Designed a solar-assisted auto rickshaw with a recharging infrastructure.[11]
4.	2011	Fuel cell and battery	A fuel cell hybrid three-wheeler has been developed to analyse the performance of fuel cell and battery capacity.[12]
5.	2014	Solar and battery	A three-wheeled vehicle model using hybrid sources has been designed and developed to improve its performance.[13]
6.	2015	Wind and battery	A wind turbine–mounted vehicle has been designed and implemented to generate 200 W at a speed of 80 km/h.[14]
7.	2016	Solar and battery	The performance analysis of a micro-controller-based PV three-wheeler has been performed to reduce the discharging time of battery.[15]
8.	2016	Battery	A detailed comparison between the lead-acid and lithium-ion battery–powered three-wheeler EVs has been established.[16]
9.	2017	Solar, ICE and battery	A solar-based plug-in hybrid three-wheeler has been developed to observe its feasibility[17]
10.	2018	Wind, solar, ICE and battery	Solar and wind energy–assisted plug-in hybrid three-wheeler has been developed to reduce the dependency on grid electricity and fuel[18]

5.4.1 Modelling of system components

A comprehensive modelling framework of the various subsystems involved in electric vehicle (EV) realization is crucial for understanding the design methodology and assessing overall system performance. Accurate component-level modelling enables effective evaluation of vehicle behavior under different operating conditions.

Vehicle dynamic modelling:

Based on Newton's laws of motion, the vehicle is represented as a mechanical load by accounting for the longitudinal forces acting along the direction of motion, as illustrated in Figure. The tractive force required to propel the vehicle is determined by the need to overcome the cumulative resistive forces encountered during motion. Accordingly, the total tractive force is expressed as:

$$F_t = \underbrace{C_{rr} M g \cos\theta}_{\text{Rolling resistance}} + \underbrace{\frac{1}{2}\rho A_d C_d v^2}_{\text{Aerodynamic drag}} + \underbrace{M g \sin\theta}_{\text{Grade resistance}} + \underbrace{Ma}_{\text{Linear inertia}} + \underbrace{F_{va}}_{\text{Rotational inertia}}$$

where *Frr* denotes rolling resistance, *Fad* represents aerodynamic drag, *Fhc* corresponds to hill-climbing resistance, *Fla* accounts for linear acceleration effects, and *Fva* signifies the force associated with rotational inertia

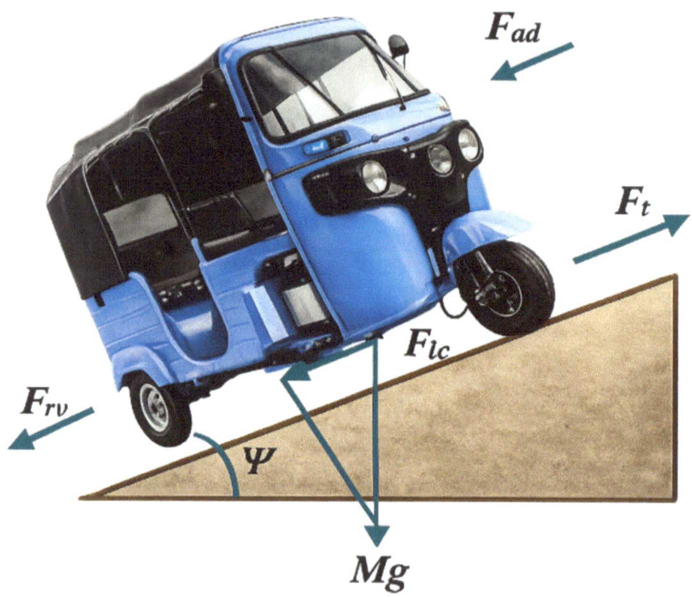

Fig.5.4: Three wheeled Vehicle

In figure, *Frr* represents the rolling resistance force, *Fad* denotes the aerodynamic drag force, *Fhc* corresponds to the hill-climbing force, *Fla* refers to the linear acceleration force, and *Fva* signifies the rotational acceleration force. The expanded mathematical expression is given as:

$$Ft = Crr\ \mathrm{Mgcos}\ C + 1/2\ \rho AdCdv^2 + \mathrm{Mgsin}\ C + Ma + Fva$$

where *Crr* is the rolling resistance coefficient, *M* is the vehicle mass (kg), *g* is the gravitational acceleration (9.81 m^2/s), denotes the road grade angle (degrees), is the air density (1.22 kg/m^3), represents the frontal area (m^2), is the aerodynamic drag coefficient, and is the vehicle speed (m/s).

The road gradient (*C*) plays a significant role in determining the dynamic response of the vehicle. To simplify the analysis, the maximum road gradient is assumed to be $C = 6°$, with the vehicle operating at a maximum speed of 15 km/h. Under these conditions, the resultant longitudinal forces acting on the vehicle generate the instantaneous tractive force required for propulsion under the considered driving cycle, as illustrated in Figure.

Using the computed instantaneous tractive force and the corresponding vehicle speed, the tractive power demand at the wheels is calculated as:

$$Pt = Ft \times v$$

Subsequently, the required motor power is obtained from the tractive power by accounting for transmission losses, which is expressed as:

$$P_m = \frac{P_t}{\eta_t}$$

where η_t represents the transmission efficiency.

Table 5.1: Various Design Concepts of a Three-Wheeled Vehicle (2005–2018)

S. No.	Year	Energy Source	Description
1	2005	Battery and ICE	Hybrid powertrain proposed as an alternative to conventional ICE vehicles; satisfactory results observed in power-assist and range-extension modes.
2	2009	Battery and ICE	Micro-hybridized Bajaj RE2 developed, showing 21% reduction in fuel consumption and emissions, with improved driver comfort.
3	2010	Solar and Battery	Solar-assisted auto rickshaw designed with supporting recharging infrastructure.

4	2011	Fuel Cell and Battery	Fuel cell hybrid three-wheeler was developed to analyse performance based on fuel cell and battery capacity.
5	2014	Solar and Battery	Hybrid-source three-wheeled vehicle model was designed and developed to improve performance.
6	2015	Wind and Battery	Wind turbine–mounted vehicle designed to generate 200 W at a speed of 80 km/h.
7	2016	Solar and Battery	Performance analysis of a microcontroller-based PV three-wheeler performed to reduce battery discharging time.
8	2016	Battery	Comparative study between lead-acid and lithium-ion battery-powered three-wheeler EVs.
9	2017	Solar, ICE, and Battery	Solar-based plug-in hybrid three-wheeler developed to assess feasibility.
10	2018	Wind, Solar, ICE, and Battery	Solar- and wind-assisted plug-in hybrid three-wheeler developed to reduce dependence on grid electricity and fuel.

5.4.2 Sizing of HESS

This section outlines a systematic design methodology for sizing the hybrid energy storage system (HESS) of a three-wheeled electric vehicle operating under the Indian Driving Cycle (IDC). The primary objective is to determine the appropriate number of lithium-ion battery cells and ultracapacitor (UC) cells required to meet vehicle power demands. Accurate sizing of the energy sources necessitates the estimation of the maximum power requirements imposed on the HESS under different operating modes defined by the IDC.

Power requirement during constant-speed operation:

During steady-state cruising, the propulsion power demand is supplied exclusively by the primary energy source, namely the lithium-ion battery. Under the IDC, the maximum constant vehicle speed is 38 km/h, at which the corresponding motor cruising power is calculated to be 1.78 kW.

Power requirement during acceleration:

The most demanding acceleration condition occurs during a transition from standstill to 22 km/h within 8 s. Under this scenario, the total power required from the HESS is estimated to be 5.25 kW. The battery is capable of delivering a maximum power of 2.46 kW; therefore, the remaining power demand of 3.1 kW is supplied by the UC module, assuming a converter efficiency of 90%.

Power requirement during road gradient operation:

For gradeability analysis, the power demand associated with road inclination is assumed to be met solely by the battery system. Considering a road gradient of 6° (equivalent to 10%) at a vehicle speed of 15 km/h, the required propulsion power is calculated to be 2.73 kW.

Power generation during deceleration:

During regenerative braking, the vehicle supplies electrical power back to the energy storage system. For a deceleration event from a maximum speed of 42 km/h to rest over a duration of 21 s, the maximum recoverable electrical power is estimated to be 1.76 kW. Based on both average and peak power demands observed across all driving modes, the sizing of the battery pack and UC bank is subsequently carried out.

Modelling of the lithium-ion battery:

An accurate lithium-ion battery model is essential for predicting battery behavior under varying operating conditions and for estimating key state variables, including the state of charge (SOC) and state of health. An equivalent electrical circuit–based battery model is illustrated in Figure. The model comprises an open-circuit voltage source , a series resistance representing ohmic losses, and a series capacitor corresponding to charge storage. Additionally, a parallel RC network is incorporated to model diffusion effects, where denotes the diffusion resistance and represents the diffusion capacitance. The parameter indicates the number of lithium-ion cells connected in series.

The terminal voltage of the battery, measured across its output terminals, is expressed as:

$$V_t = V_{OCV} - (V_1 + V_2 + V_3)$$

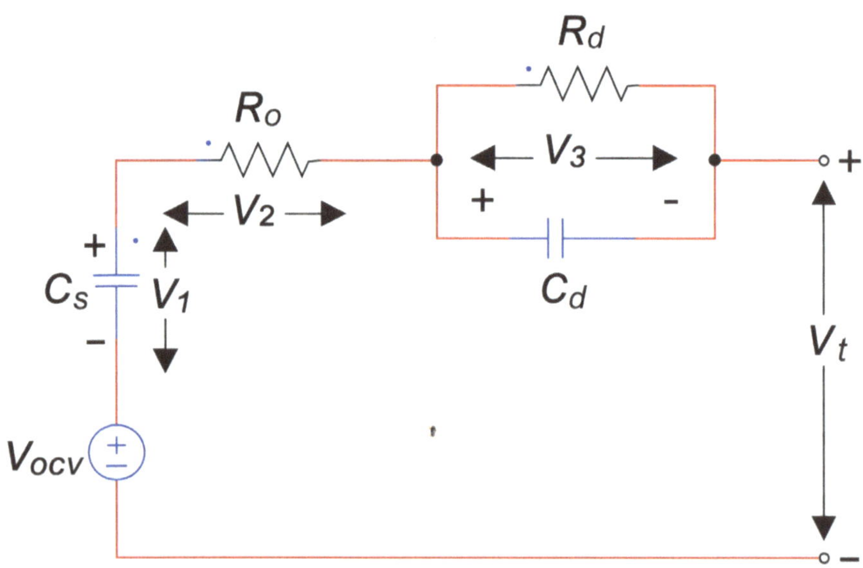

Fig.5.5: Equivalent electrical circuit–based Li-ion model.

The battery pack with a terminal voltage of 48V is directly connected to the motor drive. The battery is assumed to operate with an SOC from

30% to 100%, with an initial SOC of 80%. The energy demanded from the battery for a hybridized system is reduced to 10MJ, which is 2.78kWh. As the allowable depth of discharge (DOD) is 70%, the energy capacity required is calculated as 3.61 kWh. The specification of the Li-ion battery bank is given in Table. Hence, for a 60-km range per charge, 60 cells (4 stacks of battery with15 cells in series) are needed for 3.61 kWh of energy storage.

Range Modelling for Hybrid Vehicles (HEVs - ICE + Battery)

In electric vehicles (EVs), the driving range (distance travelled on a full charge) is determined by the total usable battery energy divided by the vehicle's energy consumption per distance. Accurately estimating the driving range of hybrid electric vehicles (HEVs) requires understanding the relationship between vehicle energy sources, powertrain configurations, and driving patterns. Range modelling draws on vehicle-specific parameters, environmental conditions, and operational cycles to predict performance under various scenarios.

Driving Cycles: Driving cycles are standardized or measured profiles of vehicle speed versus time, representing real or simulated operating conditions. They are essential inputs for range modellling as they determine the vehicle's acceleration, deceleration, and cruising phases, which in turn influence energy consumption.

Types of Driving Cycles

- Standard Cycles: Laboratory-developed profiles such as FTP-75 or WLTP for regulatory testing.
- Real-World Cycles: Derived from on-road measurements to reflect typical user behaviour.
- Custom/Test Track Cycles: Designed for specific research or validation purposes.
- Application-specific Cycles: Tailored to specialized vehicles or use cases (e.g., delivery vans, city buses).

These cycles serve as the basis for simulating energy consumption patterns in hybrid vehicles.In formal terms, the range R (in kilometres) is given by:

This formula shows that we convert the battery's energy from kilowatt-hours to watt-hours (by multiplying by 1000) and then divide by the energy use per kilometres

$$\text{Range (km)} = \frac{\text{Usable Battery Capacity (kWh)} \times 1000\ \text{Wh/kWh}}{\text{Energy Consumption (Wh/km)}}.$$

This equation is often presented in textbooks in a boxed format to highlight its importance. It explicitly includes unit annotations: "kWh" for battery capacity and "Wh/km" for consumption, making unit consistency clear (since 1 kWh = 1000 Wh). In words, the total available energy (in Wh) divided by the energy used per km yields the distance (km) the EV can travel.

Terms and Units:

The total distance an EV can travel on one full charge. By convention, this is the distance when the battery's usable energy is fully expended. It has units of kilometres (km). Usable Battery Capacity (kWh): The amount of electrical energy available from the battery (in kilowatt-hours). Importantly, this is the usable portion of the battery's total (gross) capacity, excluding any buffer reserved to prolong battery life. For example, a battery pack might have a gross capacity of 32.6 kWh but only 28.9 kWh is usable. Physically, 1 kWh is a unit of energy equal to the output of 1 kilowatt (kW) of power sustained for 1 hour. Because 1 kW = 1000 W, one kilowatt-hour equals 1000 watt-hours. In the formula, we multiply by 1000 to convert kWh to Wh so that unit's match. Energy Consumption (Wh/km): The electrical energy required to drive one kilometre. This metric (often determined by official test procedures) is expressed in watt-hours per kilometre. It represents the EV's efficiency

This aligns with the notion that distance = energy / (energy per distance).

Numerical :

Usable Battery Capacity = 30 kWh; Energy Consumption = 150 Wh/km.

1. Convert battery capacity to watt-hours:
   30 kWh=30×1000 Wh=30000 Wh.
2. Divide by the consumption rate:
   Range=150 Wh.
3. Thus, an EV with a 30-kWh usable battery and a consumption of 150 Wh/km has a theoretical range of 200 km. This follows directly from the formula:

Each step respects the units and physical meaning: the battery provides 30 kWh (30,000 Wh) of energy, and at 150 Wh per km the vehicle uses 150 Wh for each kilometer, so it can go 30000/150 = 200 kilometres on that energy.

Interpretation: More available energy or lower consumption yields greater range. In practical terms, manufacturers use the usable kWh of the battery and the certified Wh/km efficiency to compute range. This calculation underlies the range values quoted on vehicle labels and in test procedures.

5.5 Architecture: Hybrid Vehicle Architecture

Hybrid Electric Vehicles (HEVs) are designed using different powertrain architectures that define how the internal combustion engine (ICE) and the electric motor deliver energy to the wheels. The architecture determines performance, efficiency, and suitability for different driving scenarios.

1. **Series Hybrid Architecture-**The ICE is not mechanically connected to the wheels. The engine drives a generator, which produces electricity to either power the motor directly or charge the battery. The motor alone provides traction. Advantages: Simple mechanical layout, good for urban stop-and-go traffic. Example: Chevrolet Volt (in extended EV mode).

2. **Parallel Hybrid Architecture-**Both the ICE and electric motor are mechanically connected to the wheels. Either power source can propel the vehicle independently or together. Advantages: High efficiency at highway speeds, flexible operation. Example: Honda Insight.

3. **Series-Parallel (Power-Split) Architecture- Combines** elements of both series and parallel systems. A planetary gearset or power-split device allows both the engine and motor to drive the wheels or charge the battery. Advantages: Optimal balance between performance and efficiency. Example: Toyota Prius, Hyundai Ioniq.

Energy Sources in Hybrid Vehicles: From ICE: Provides steady, high-power energy for long-distance cruising. From Battery: Supplies instantaneous torque for acceleration and low-speed operation. From Both: Used during high-demand situations such as overtaking, hill climbing, or rapid acceleration.

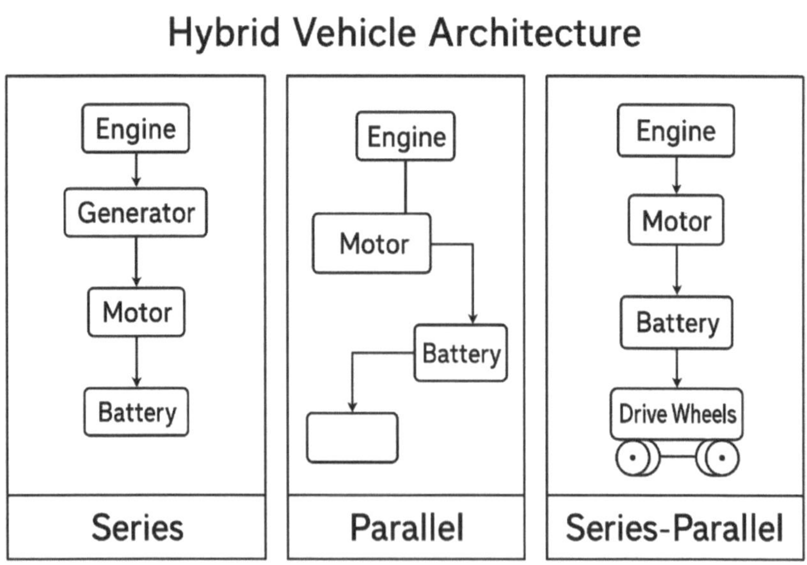

Fig.5.6: Different powertrain architectures

5.6 Power Split and Energy Management in Hybrid Vehicles

The performance of hybrid electric vehicles (HEVs) relies heavily on their power split logic and energy management algorithm. These control strategies determine how the energy from the internal combustion engine (ICE), the battery, or both is allocated to the drivetrain. The State of Charge (SOC) bounds are maintained to ensure battery longevity and prevent deep discharge or overcharging. The model architecture typically consists of several core blocks. The ICE fuel consumption map characterizes the engine's efficiency under different load and speed conditions. The battery energy flow model governs charging and discharging patterns in coordination with regenerative braking. Additionally, generator and motor dynamics models simulate torque production, rotational speed, and power transfer. Together, these elements allow the system to balance efficiency, performance, and emissions.

5.7 Fuel Cell Electric Vehicle (FCEV) Range Modelling

In fuel cell electric vehicles, the primary energy source is a hydrogen fuel cell, supported by a small battery for transient loads and regenerative braking storage. The driving range of an FCEV is influenced by several parameters, including hydrogen tank capacity (measured in kilograms), fuel cell efficiency, and vehicle energy consumption per kilometre (Wh/km).

The range can be estimated using the formula

$$Pange = \frac{\text{Hydrogen Capacity (kg)} \times \text{LHV} \times \eta_{FC}}{\text{Vehicle Energy Consumption (Wh/km)}}$$

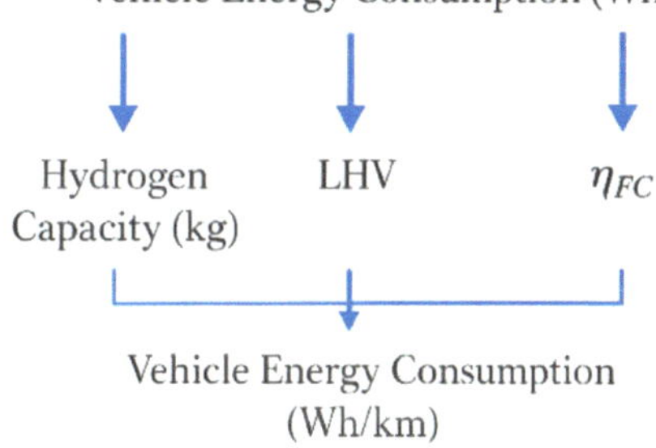

where the Lower Heating Value (LHV) of hydrogen is approximately 33.3 kWh/kg, and ηFC denotes fuel cell efficiency. For example, a vehicle with a 5 kg hydrogen tank, combined with typical efficiency and energy consumption values, can achieve a range of approximately 350–400 km.

Solar-Powered Vehicle Range Modelling

Solar-powered vehicles use photovoltaic panels to convert solar irradiance into electrical energy, which is stored in a battery and used for propulsion. The key factors influencing solar energy capture are solar panel area, panel efficiency, and irradiance (W/m^2).

The instantaneous power output from the solar panels can be expressed as

$$P_{\text{solar}} = \text{Area} \times \text{Irradiance} \times \text{Efficiency}$$

Since solar energy availability depends on geographic location, time of day, and vehicle exposure, solar panels generally serve as a supplementary energy source rather than the primary power provider—except in specialized cases such as solar racing vehicles or lightweight low-speed urban commuters.

Case Studies: Range Modelling

Case Study: 2-Wheeler Electric Scooter (India)

Parameter	Value
Battery	2 kWh (LFP)
Efficiency	45 Wh/km
Regeneration	~5% recovery
Range (IDC)	~48–50 km
Charging Time	3-4 hrs (AC)
Driving Cycle Used	WMTC or MIDC

Case Study : 3-Wheeler E-Rickshaw

Parameter	Value
Battery	4.5 kWh
Energy Consumption	70–80 Wh/km
Operating Range	60–70 km
Load Factor	High (3-4 passengers + cargo)
Driving Conditions	Stop-and-go, urban

Case Study : 4-Wheeler Hybrid SUV

Parameter	Value
Engine	1.5L Petrol + Motor
Battery Capacity	1.3 kWh (NiMH or Li-ion)
Drive Type	Series-Parallel Hybrid
Fuel Tank	40 Litre
Range (combined)	750–800 km
Electric-only Range	2–4 km (PHEV = 40–60 km)
Driving Cycle	WLTP, Urban & Highway mix

Table 5.2: Comparison of EV Range Modelling Approaches

Type	Energy Source	Range Equation	Influencing Factors
BEV	Battery	kWh/ Wh/km	SOC, driving cycle, terrain
HEV	ICE + Battery	Fuel map + Battery	Power split strategy, SOC profile
FCEV	H_2 Fuel Cell + Battery	kg×LHV×ηWh/ Energy consumption (Wh/km)	H_2 tank, FC efficiency

Solar EV	Solar Panel + Battery	Irradiance dependent	Sunlight, panel size, exposure

Tools for Range Modelling -MATLAB/Simulink – Full EV & HEV simulations, Autonomies (by Argonne National Lab), ADVISOR (Advanced Vehicle Simulator), GT-Suite, AVL CRUISE – Professional software

Rememberence

- *Battery capacity (kWh)*
- *Energy consumption (Wh/km)*
- *Regenerative braking recovery*
- *Driving cycle type*

Modelling the range of hybrid, electric, and alternative-energy vehicles helps optimize design and deployment. With diverse driving cycles and vehicle configurations, range modelling becomes a critical component in the sustainable transportation ecosystem.

Multiple Choice Questions

1. **Vehicle range in the context of Hybrid Electric Vehicles (HEVs) is defined as:**

   a) Maximum speed a vehicle can achieve on full fuel
   b) Maximum distance travelled per unit fuel consumption
   c) Maximum distance travelled on full fuel/battery energy replenishment
   d) Minimum distance before refuelling

   Answer: c

2. **Which factor is NOT a primary determinant of HEV range?**

   a) Powertrain configuration
   b) Driving cycle type
   c) Ambient environmental conditions
   d) Paint colour of the vehicle

   Answer: d

3. **Driving cycles are important in HEV range modelling because they:**

   a) Reduce battery degradation rate
   b) Provide standardized, repeatable speed-time patterns
   c) Directly measure hydrogen tank pressure
   d) Eliminate the need for simulations

   Answer: b

4. **FTP-75 is mainly used for:**

   a) High-speed track testing
   b) Emission certification and fuel economy testing in the USA
   c) Measuring regenerative braking efficiency only
   d) Testing of heavy-duty engines exclusively

   Answer: b

5. In the FTP-75 driving cycle, the average speed is approximately:

a) 21.2 mph
b) 56.7 mph
c) 34.12 km/h
d) Both a and c

Answer: d

6. The Indian Driving Cycle (IDC) is best classified as:

a) Standard cycle
b) Real-world cycle
c) Custom/test-track cycle
d) Application-specific cycle

Answer: c

7. Which type of driving cycle is derived from on-road measurements?

a) Real-world cycle
b) Standard cycle
c) Test-track cycle
d) Application-specific cycle

Answer: a

8. WLTP stands for:

a) Worldwide Light Transport Protocol
b) Worldwide Harmonized Light Vehicles Test Procedure
c) World Light Truck Performance
d) None of the above

Answer: b

9. The formula for BEV range is:

a) Range = (Usable Battery Capacity in kWh) ÷ (Energy Consumption in Wh/km)
b) Range = (Usable Battery Capacity × 1000) ÷ (Energy Consumption)
c) Range = Fuel Tank Capacity ÷ Fuel Consumption
d) Range = Distance × Time

Answer: b

10. A BEV has 40 kWh usable capacity and consumes 200 Wh/km. The theoretical range is:

a) 150 km
b) 200 km
c) 250 km
d) 300 km

Answer: c

(Calculation: 40 × 1000 ÷ 200 = 20000 ÷ 200 = 200 → Wait, correction: 40 × 1000 = 40000 Wh; 40000 ÷ 200 = 200 km, so correct answer is b)

Correct Answer: b

11. In FCEVs, range can be estimated using:

a) Range = (H_2 mass × LHV × Fuel Cell Efficiency) ÷ Energy Consumption
b) Range = Battery capacity ÷ Power output
c) Range = LHV × Hydrogen tank volume
d) Range = Energy × Speed

Answer: a

12. If an FCEV has a 5 kg hydrogen tank, LHV of hydrogen is 33.3 kWh/kg, η = 55%, and consumption is 180 Wh/km, the range is:

a) ~400 km
b) ~500 km
c) ~300 km
d) ~250 km

Answer: a

(Calculation: 5 × 33.3 × 0.55 = 91.575 kWh; in Wh → 91575 Wh; ÷ 180 Wh/km ≈ 508.75 → Wait, that's ~509 km, so correct answer is b)

Correct Answer: b

13. In series hybrid architecture, the ICE:

a) Drives the wheels directly
b) Is not connected mechanically to the wheels

c) Only works in parallel with the motor

d) Is used for regenerative braking

Answer: b

14. The parallel hybrid is most efficient for:

a) Urban stop-and-go traffic

b) High-speed highway driving

c) Off-road driving only

d) Stationary power generation

Answer: b

15. In series-parallel hybrids, the power split device is usually:

a) Belt drive

b) Planetary gearset

c) CVT gearbox

d) Hydraulic coupling

Answer: b

16. The main purpose of SOC bounds in HEV control is to:

a) Increase motor torque

b) Prevent deep discharge or overcharge of the battery

c) Increase fuel tank capacity

d) Enhance regenerative braking speed

Answer: b

17. A 2-wheeler EV in India with a 2 kWh battery, efficiency of 45 Wh/km, and ~5% regenerative recovery will have an IDC range of approximately:

a) 40–42 km

b) 45–47 km

c) 48–50 km

d) 52–55 km

Answer: c

18. A 3-wheeler e-rickshaw with 4.5 kWh battery and energy consumption of 80 Wh/km will have a theoretical range of:

a) 50 km

b) 56 km

c) 60 km

d) 65 km

Answer: c

(Calculation: 4.5 × 1000 ÷ 80 = 4500 ÷ 80 = 56.25 km → Wait, correction: 4500 ÷ 80 = 56.25, so correct answer is b)

Correct Answer: b

India-GTU Questions

Give the Differences Between BEV, HEV and FCEV. chapt5pg147 **03**

Chapter 06

Energy Management Strategies

6 ENERGY MANAGEMENT STRATEGIES

6.1 Introduction to EMS

Energy Management Strategies (EMS) are decision-making frameworks designed to control and optimize the flow of power between different energy sources, such as the battery, engine, and motor. Their primary goals include improving fuel economy, reducing harmful emissions, extending battery life, and enhancing overall vehicle performance. These strategies are crucial in ensuring that the right energy source is used at the right time, depending on driving conditions and vehicle demands. EMS plays a vital role in various types of electric mobility systems, including Battery Electric Vehicles (BEVs), Hybrid Electric Vehicles (HEVs), and Plug-in Hybrid Electric Vehicles (PHEVs), making them a key element in modern sustainable transportation. Hybrid Electric Vehicles (HEVs) and Electric Vehicles (EVs) utilize various Energy Management Strategies (EMSs) to optimize performance, fuel economy, and battery life. These strategies dictate how power is allocated between different energy sources (engine, battery, motor) to meet driving demands while minimizing undesirable effects like fuel consumption and emissions. Here's a breakdown of the key strategies and their implications.

Hybrid electric vehicles (HEVs) are emerging as a key transitional technology in the automotive sector as the industry moves toward low-emission transportation. Unlike conventional vehicles, HEVs rely on a combination of an internal combustion engine and an electric motor powered by a battery system. To ensure efficient operation, these vehicles depend on a well-designed energy management strategy (EMS) that dynamically determines how power is split between the engine and the electric motor while maintaining the battery's state of charge within safe and optimal limits.

An effective EMS must respond rapidly to varying driver demands, road conditions, and traffic patterns. At the same time, it should minimize fuel consumption and emissions by intelligently predicting future power requirements, battery charge–discharge behaviour, and engine efficiency. This real-time decision-making becomes especially challenging under uncertain and highly dynamic driving conditions, such as urban stop-and-go traffic or sudden acceleration events.

As HEVs gain wider adoption, EMS technologies are evolving from traditional control methods to more intelligent and predictive systems. Existing EMS approaches can be broadly classified into rule-based strategies and optimization-based strategies. Rule-based methods rely on predefined thresholds and heuristics, making them simple and reliable but limited in adaptability. In contrast, optimization-based methods aim to achieve global or near-optimal energy usage by solving mathematical models, offering better efficiency at the cost of higher computational complexity.

6.1.1 Evolution of EMS Technologies

The development of EMS can be visualized through a technology roadmap that reflects increasing intelligence and autonomy. Look at the anatomy below,

Conventional EMS

↓

Rule-Based Control

↓

Optimization-Based Control

↓

Machine Learning–Based EMS

↓

Reinforcement & Deep Learning EMS

Modern EMS research focuses on integrating advanced algorithms such as reinforcement learning and deep neural networks. These methods allow the system to learn from driving data, adapt to individual driver behaviour, and improve decision-making over time without relying solely on predefined rules.

6.1.2 Enabling Technologies

The advancement of energy management systems (EMS) for hybrid electric vehicles is driven by the integration of several enabling technologies that enhance intelligence, adaptability, and efficiency. Machine learning techniques allow EMS to learn from historical and real-time driving data, enabling adaptive control and accurate prediction of energy demand, battery state of charge, and fuel consumption. Cloud computing complements this capability by providing high computational power and storage, supporting large-scale data analysis, continuous model training, and over-the-air updates of EMS algorithms. Computer vision further strengthens EMS performance by offering real-time awareness of the driving environment, such as traffic conditions, road gradients, and obstacle detection, which helps the system anticipate power requirements more effectively. In addition, swarm intelligence enables cooperative learning among multiple vehicles, allowing shared experiences and collective optimization of energy usage. While these technologies significantly improve EMS performance, they also introduce challenges related to computational complexity, cybersecurity, real-time deployment, and overall system reliability.

6.1.3 Example: EMS Decision Scenario

In an urban driving scenario characterized by low vehicle speeds and frequent braking, the EMS prioritizes the use of the electric motor to reduce fuel consumption and emissions, while simultaneously recovering energy through regenerative braking to recharge the battery. In contrast, during highway driving conditions with relatively constant speeds and higher power demand, the EMS shifts a greater portion of the load to the internal combustion engine, maintaining the battery state of charge within an optimal range to ensure long-term efficiency and system stability.

6.1.4 Industry Context and Impact

For over a century, the automotive industry has been dominated by internal combustion engines. However, increasing concerns about climate change, air pollution, and energy security have accelerated the transition toward electrified transportation. Many countries have introduced regulations or ban on vehicles powered solely by internal combustion engines due to their environmental impact.

Although fully electric vehicles represent the long-term objective, widespread adoption is currently constrained by battery cost, charging infrastructure, and driving range limitations. In this context, HEVs provide a practical and effective intermediate solution by combining the benefits of electric propulsion with the reliability of conventional engines.

Key Insight

The continued advancement of EMS technologies will play a decisive role in enhancing fuel economy, reducing emissions, and improving driver experience. By leveraging intelligent algorithms and connected technologies, next-generation EMS systems are positioned to make HEVs smarter, cleaner, and more responsive—a massive upgrade for sustainable mobility.

Inherent in HEVs is their energy management strategy (EMS), which sets the conditions for when the vehicle is powered by the engine versus the battery/motor assembly. The EMS plays a key role in ensuring the vehicle operates under optimal conditions to minimize fuel consumption. The goal of this review paper is to provide a technology roadmap for the EMSs used in HEVs. It first discusses the significance and current state of the EMSs in modern HEVs. It then develops a technology roadmap for future EMSs based on market trends. This paper then presents a comprehensive review of the technologies driving the evolution of EMSs, including machine learning, computer vision, cloud computing, and swarm technology. These technologies are discussed in relation to EMSs, with a particular focus on how their advances affect overall fuel and energy consumption. HEVs are expected to play a critical role in the future of the automotive industry, serving as a bridge between current internal combustion engines and fully electric vehicles. Given the limited readiness of global charging infrastructure and the higher build costs of EVs, HEVs represent the best option in the near term for certain geographical locations and user groups. HEVs offer a cleaner alternative to traditional vehicles while mitigating common issues with electric vehicles, such as range limitations and charging times. The benefits of HEVs rely heavily on having an effective EMS to optimize the vehicle's drive parameters. This study provided a summary of current and future EMSs for HEVs, highlighting the reliance on rule-based and optimization-based strategies in existing systems. Future EMSs are expected to incorporate technologies from emerging fields, including machine learning, computer vision, cloud computing, and swarm technology. The paper presents a roadmap for EMS development, predicting advancements that will enhance optimization and decrease vehicle fuel consumption, potentially approaching the MPGe of electric vehicles. This study indicates that industry stakeholders should invest in integrating advanced ML algorithms, computer vision, swarm technology, and cloud computing to enhance EMS capabilities. Collaboration with technology providers specializing in these areas and predictive analytics can accelerate the development of next-generation EMSs. Additionally, companies should advocate for and contribute to the development of charging infrastructure to support the transition to fully electric vehicles in the longer term. Further research should include longitudinal studies to assess the real-world performance of advanced EMSs and their impact on fuel consumption and vehicle efficiency, cross-disciplinary research to explore synergies between EMS technologies and other automotive innovations, and investigations into how different driving patterns and user behaviours influence the effectiveness of various EMS strategies. As EMS technology evolves, HEVs will increasingly bridge the gap until the automotive industry transitions fully to electric vehicles, achieving greater fuel efficiency and environmental benefits.

6.2 Energy Management concept for HEV GTUexam-3170923-SUM2025

The Energy Management System (EMS) in a Hybrid Electric Vehicle (HEV) is the control logic that orchestrates the power flow between its multiple energy sources, typically an internal combustion engine (ICE) and a battery-electric motor system. The core concept is to determine the most efficient power split in real-time to minimize fuel consumption and emissions while meeting the driver's performance demands.

The EMS handles several critical functions:

Deciding the power source: The system determines whether to use the engine, the electric motor, or a combination of both to propel the vehicle.

Engine operation: It manages when to turn the engine on and off and keeps its operation within the highest-efficiency regions of its performance map.

Regenerative braking: The EMS captures kinetic energy during deceleration and braking, converting it back into electricity to recharge the battery.

Battery state of charge (SOC) management: The system maintains the battery's charge within a healthy range to maximize battery life and ensure power is available when needed

6.2.1 Hybrid Topologies

As illustrated in the figure, hybrid electric vehicles (HEVs) are commonly classified according to their drivetrain layout into series, parallel, and power-split configurations, depending on how the internal combustion engine interacts with the electric motor and battery. It is worth noting that a basic level of hybridization is present even in conventional vehicles, where an electric motor powered by the battery is used solely for engine starting. However, in such systems, the electric unit does not contribute to vehicle propulsion.

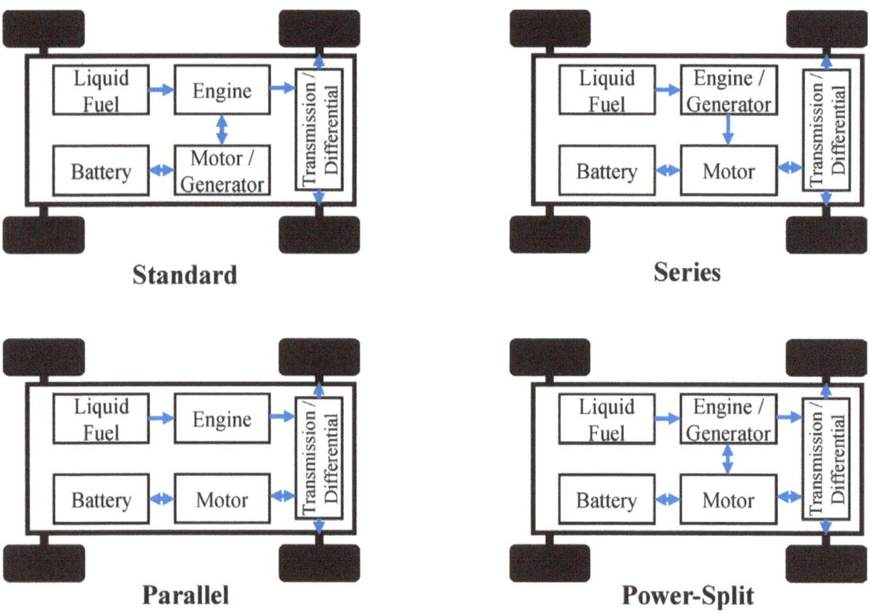

Fig.6.1: Drivetrain layout topologies

A series hybrid electric vehicle (HEV) represents the simplest hybrid configuration and operates similarly to a battery electric vehicle, with the wheels driven exclusively by an electric motor. An internal combustion engine coupled to a generator supplies electrical energy to either recharge the battery or directly power the motor. The elimination of a mechanical transmission reduces system weight, and because the engine is decoupled from driving conditions, it can operate at optimal efficiency. However, this architecture requires relatively large components, including the battery, engine, and auxiliary systems, increasing overall cost. In addition, multiple energy conversions—from mechanical to electrical and back—introduce efficiency losses.

In parallel HEVs, both the engine and the electric motor are mechanically connected to the drivetrain and can independently or jointly propel the vehicle. The energy management system (EMS) selects the most efficient power source based on operating conditions. Typically, the motor dominates at low speeds to improve fuel economy, while the engine provides propulsion at higher speeds and under greater loads. Compared to series HEVs, parallel systems require a smaller battery pack and avoid losses associated with repeated energy conversions.

Power-split HEVs combine the characteristics of series and parallel architectures to achieve higher efficiency across varied driving conditions. In this configuration, the engine, generator, and motor are decoupled and coordinated through a planetary gear set, enabling power flow through both mechanical and electrical paths. This arrangement allows flexible control of power distribution and improved engine operating efficiency.

HEVs are further classified as mild, full, and plug-in hybrids. Mild hybrids, typically based on a parallel architecture, shut down the engine during idling or braking but do not support electric-only propulsion. In contrast, full and plug-in hybrids—generally employing power-split architectures—enable electric propulsion. Both recharge the battery using an onboard generator, while plug-in hybrids additionally allow charging from the electrical grid. This study focuses on energy management strategies in full and plug-in hybrids, which are inherently more complex than those used in mild hybrid systems.

The energy management system plays a critical role in determining hybrid vehicle fuel efficiency. As shown in the figure, excessive reliance on the internal combustion engine can offset the benefits of hybridization, particularly due to the additional vehicle mass, potentially resulting in higher fuel consumption than that of a conventional vehicle. Fuel economy is commonly evaluated using miles per gallon (MPG). When an Energy Management System (EMS) effectively predicts power demand and strategically restricts engine operation, hybrid electric vehicles (HEVs) can attain significantly higher efficiency. This efficiency is often reported as miles per gallon equivalent (MPGe), and under optimized operating conditions, HEVs may even surpass the efficiency of fully electric vehicles.

MPGe quantifies how efficiently a vehicle utilizes energy by expressing the distance travelled using an energy amount equivalent to one gallon of gasoline. It serves as a standardized metric that enables direct comparison between conventional, hybrid, and electric powertrains and is defined using the following relationship:

$$\text{MPGe} = \frac{\text{Distance Travelled [miles]}}{\text{Energy Consumed [kWhr]}} \times \text{Energy Content of Gasoline [kWhr/gallon]}$$

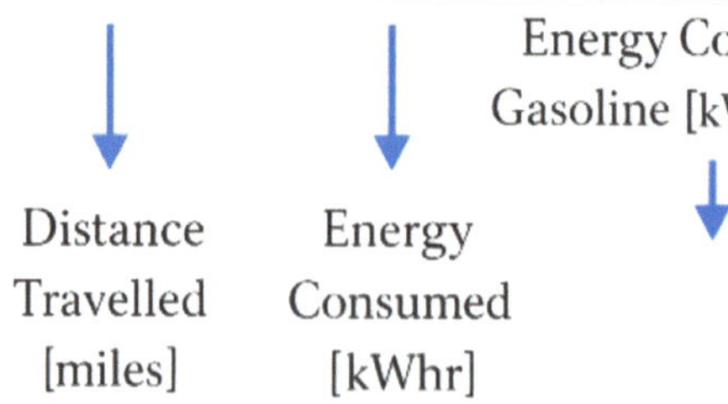

The comparatively higher MPGe of a hybrid electric vehicle (HEV), when evaluated against a similar electric vehicle (EV), can be attributed to the HEV's smaller battery pack. This reduced battery capacity lowers the overall vehicle mass, which in turn decreases the power required for propulsion, leading to reduced energy consumption during locomotion.

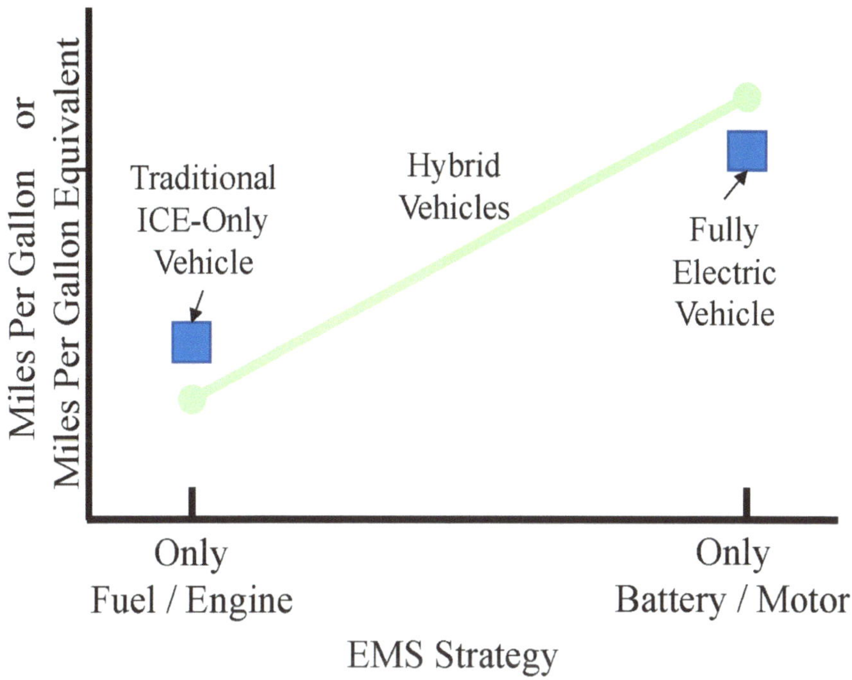

Fig.6.2: Qualitative comparison of miles per gallon

The figure presents a qualitative comparison of fuel efficiency, expressed as miles per gallon (MPG) and miles per gallon equivalent (MPGe), across an internal combustion engine vehicle, an electric vehicle, and a hybrid electric vehicle (HEV). Developing an optimal EMS is challenging, primarily due to the need to balance performance with fuel economy. The EMS must quickly respond to driver demands while optimizing energy use, which becomes difficult in unpredictable driving conditions. Ideally, it would accurately predict future battery charge rates and fuel consumption, allowing for real-time adjustments to the powertrain. However, these predictions are complex and often imprecise, making it hard for the EMS to anticipate future driving conditions and optimize the power split effectively.

6.2.2 Current Energy Management Strategies

Rule-Based The most basic EMS is a rule-based EMS, which predetermines the operating states of a hybrid system using predefined rules to manage how the system functions. These rules are built from heuristics, intuition, human experience, and mathematical models tailored to specific scenarios. The EMS distributes power between the engine and battery, ensuring both operate within their most efficient ranges. There are two primary types of rule-based EMSs. The first is deterministic, which controls vehicle components based on specific driving demands and system limits. This method relies on established parameters for each component, combined with engineering experience and research data. It adjusts power distribution and component status dynamically, responding to changes in driving conditions and system requirements. The second type is a fuzzy logic-based EMS, which enhances the standard rule-based system by incorporating fuzzy logic. Instead of focusing solely on state variables like power demand and vehicle speed, this system also accounts for the rate of change in these variables, offering a more nuanced approach to decision-making. While rule-based strategies are effective and relatively easy to implement, they are constrained by specific operating conditions and lack adaptability to varying driving cycles. These controllers are popular due to their simplicity and real-time application suitability, but they often rely on basic engineering intuition, such as the charge-depleting/charge-sustaining strategy, which is less efficient than more advanced approaches. Current rule-based EMSs struggle to account for variations in trip lengths or traffic conditions without employing complex driving pattern recognition systems.

6.2.3 Optimization-Based Strategies GTUexam-3170923 SUM2025,WIN2024

*Optimization-Based Strategies-*The goal of an optimization strategy is to minimize a cost function, which typically includes emissions, fuel consumption, and torque requirements. Unlike rule-based strategies, optimization strategies do not directly manage real-time energy usage. Instead, they derive real-time control decisions from the vehicle's cost function. These algorithms typically calculate the optimal power split between the engine and motor for a specific drive cycle. However, the solutions they produce are optimal only for that particular cycle, which is usually represented as a set of vehicle speed points over time. Consequently, they may not be optimal or even charge-sustaining for other driving cycles. Without predicting future driving conditions in real time, it is impossible to directly apply these control laws during vehicle operation. Optimization-based EMS can be classified into two main types. The first is global optimization, which focuses on minimizing the cost of the entire driving condition and applies optimal control theory to achieve a global solution. The second is instantaneous optimization, which aims to minimize fuel consumption and other parameters at each moment of the drive cycle. For global optimization to perform at its best, it requires extensive data about the vehicle, such as battery SOC, driving conditions, driver behaviour, and route information. However, due to limitations in computational methods like linear programming, dynamic programming, and genetic algorithms, global optimization is not feasible for real-time control. Instead, it is typically optimized for a predetermined drive cycle. While dynamic programming (DP) can compute multi-stage optimization decisions when the entire driving cycle is known, it is too computationally intensive for real-time application. Instantaneous optimization strategies, such as model predictive control (MPC) and the equivalent consumption minimization strategy (ECMS), address some of these limitations. MPC predicts future power demand using real-time driving data and adjusts the power distribution between the engine and battery to minimize fuel consumption. ECMS, on the other hand, simplifies the dynamic optimization problem by minimizing equivalent fuel consumption at each time step, turning it into an instantaneous optimization problem. An alternative to optimization is a dynamic rule-based strategy, where the rule set is continuously updated based on changing driving conditions. For example, Basma et al. studied a dynamic rule-based EMS that incorporated elements of dynamic programming. Their study proposed a comprehensive methodology for designing EMS in HEVs to achieve near-optimal consumption results.

The study reported that for travel distances of up to 120 km, the dynamic rule-based Energy Management System (EMS) achieved a 15% reduction in fuel consumption compared with a conventional rule-based approach. While the absolute decrease in fuel usage was modest, even marginal improvements can lead to meaningful enhancements in overall vehicle performance. The proposed control strategy effectively combines the performance benefits of global optimization methods with the practicality of real-time implementation.

Limitations: Although both rule-based and optimization-based approaches are widely applied in hybrid energy management, they predominantly rely on instantaneous inputs such as battery state of charge, vehicle speed, and immediate power demand, as illustrated in the figure. Consequently, the inputs used by current Energy Management Systems (EMSs) are largely reactive, as they respond only to present operating conditions. Several influential factors remain unaddressed, most importantly, these approaches fail to incorporate predictions of the vehicle's future energy demand. In addition, they generally ignore external influences—such as traffic patterns and weather conditions—that can have a substantial impact on overall vehicle efficiency. To highlight these shortcomings, four representative scenarios are examined below.

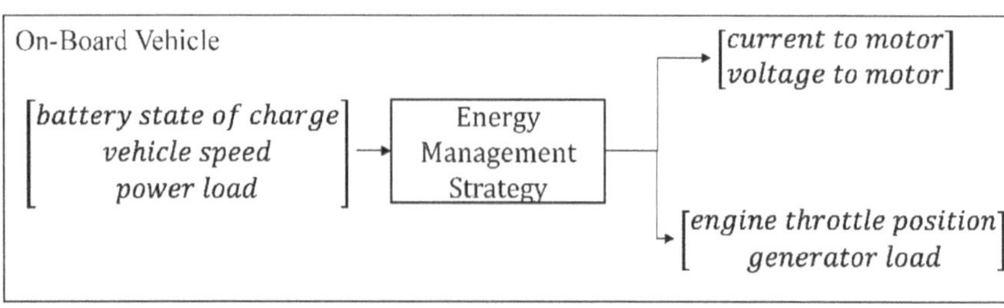

Fig.6.3: State diagram for the current EMS for HEV.

Case 1:

Consider a vehicle traveling at highway speeds that is about to enter an urban environment characterized by frequent braking events. Conventional EMS approaches tend to increase engine output during highway cruising to maintain a high battery state of charge. However, once the vehicle transitions to city driving, a fully charged battery cannot accept additional energy from regenerative braking, resulting in lost recovery potential. A predictive EMS that anticipates the upcoming urban conditions could regulate the battery charge level in advance, ensuring sufficient capacity to capture regenerative energy and thereby improving overall efficiency.

Case 2:

In scenarios involving extended low-speed driving, electric propulsion is generally preferred. In series or power-split hybrid architectures, the internal combustion engine may still be activated to recharge the battery. If the vehicle is operating as a plug-in hybrid and is approaching its destination or a charging location, engine-based charging becomes inefficient compared to grid charging. A predictive EMS that accounts for proximity to charging infrastructure could avoid unnecessary engine operation, conserving fuel and deferring battery replenishment to external electrical sources.

Case 3:

A vehicle operating at moderate speeds may initially rely on electric propulsion but later encounter severe traffic congestion, resulting in prolonged idling. Without anticipation of such conditions, the battery may become depleted during the idle period, requiring the engine to engage and increasing fuel consumption and emissions. A predictive EMS with knowledge of upcoming traffic conditions could primitively utilize the engine to maintain an adequate battery charge, thereby reducing inefficiencies during extended idling.

Case 4:

When approaching a sustained uphill gradient, operating the engine concurrently with hybrid propulsion during the climb can lead to suboptimal efficiency and increased system load. If the EMS has prior knowledge of the terrain, it can schedule higher engine loading before reaching the incline to store additional energy in the battery. This stored energy can then be used during the ascent, enabling more efficient hill climbing and improved overall energy utilization.

Across all four scenarios, enhanced vehicle efficiency is achievable through an EMS that incorporates predictive capabilities, integrates driver behaviour patterns, and accounts for external factors to anticipate future power demands.

6.3 Solar and grid-connected charging station

A schematic diagram of a solar and grid-connected charging station shows the flow of power from multiple sources to electric vehicles (EVs), with a battery storage system balancing supply and demand. A smart Energy Management System (EMS) controls how energy is used to prioritize solar power, reduce costs, and ensure reliable charging.

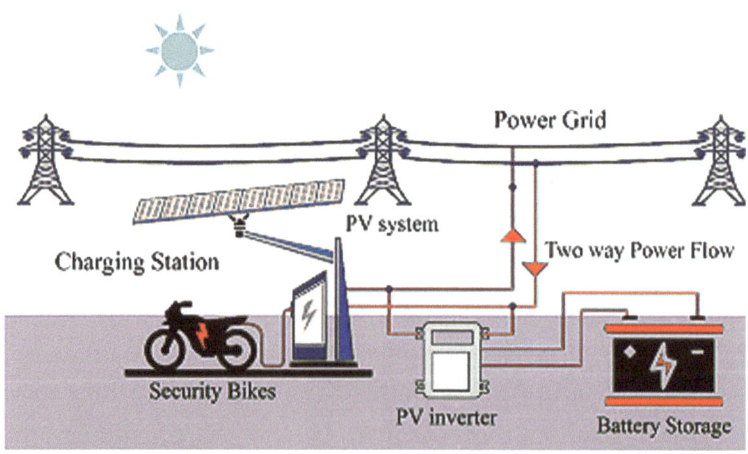

Fig.6.4: solar and grid-connected charging station

6.3.1 Technology Roadmap

Given the issues discussed in the previous section, the EMS will evolve to better optimize fuel consumption and vehicle performance. Figure displays a projected roadmap for the EMS for HEVs based on the evolution of the underlying technologies. In particular, the current rule-based and optimization-based EMSs will be replaced with more advanced versions that can adapt to a driver's habits. Over the long term, EMS will evolve to include numerous external factors, including traffic and weather. Each of these stages of EMS technology will be discussed in the subsequent sections.

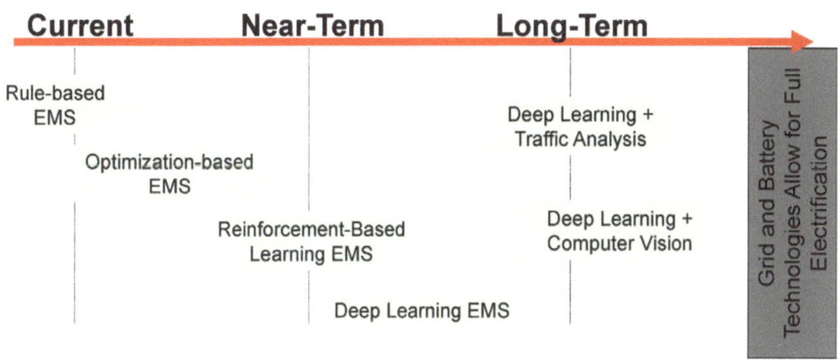

Fig.6.5: Technology roadmap for the evolution of EMSs for HEVs.

"Long-term" in this roadmap is loosely defined as being 20 years out. By the end of that period, HEVs may become obsolete to fully electric vehicles with new battery chemistries. These batteries will eliminate the weight and cost issues associated with lithium-ion batteries. Further, roadmaps past 20 years for a rapidly changing field carry significant uncertainty.

A summary of each EMS is given in table, including their positive and negative characteristics. The near- and mid/long-term EMSs are discussed in detail in the followings.

Table 6.1: Summary of EMS

EMS	Positives	Negatives
Rule-Based	Simple with minimal computational power. Uses existing technology.	Does not account for future driver needs or individual driving habits.

Optimization-Based	Improved fuel efficiency compared to rule-based EMS.	More computationally intensive than rule-based EMS. Does not account for future driver needs or individual driving habits.
Reinforcement-Based Learning	Improved fuel efficiency from predicting future states. Behaviour tailored to individual driver.	Computationally intensive and difficult to achieve real-time control. Uncertainty in future driver need predictions.
Deep Learning-Based	Improved fuel efficiency from predicting future states with reducing uncertainty in future state predictions.	Computationally intensive and difficult to achieve real-time control.

6.3.2 Current Scenarios:

Near-Term EMS-Current rule-based and optimization-based EMSs do not account for individual driving habits. However, in the near term, EMSs will be able to adapt to unique driving conditions, as shown in fig. In particular, reinforcement learning (RL)-based EMSs are one of the newest ways to manage energy. This strategy learns from historical data and uses the previous driving data for learning and application. An RL-based EMS can be implemented to derive the optimal control policy by using inputs such as vehicle speed, driver power demand, and SOC to determine the engine power. This approach adapts to unpredictable driving cycles and achieves lower fuel consumption compared to dynamic programming.

Additionally, the EMS can take in data from other vehicles in the area, especially those that are ahead of the current vehicle. In doing so, the EMS can receive data that may not necessarily be available through traffic data or from onboard sensors. For example, the system would be able to identify the instant that the vehicle in front of it starts to brake. Moreover, the EMS could detect when additional, unpredicted loads may be placed on the vehicle, such as a wind gust or a change in road grade. The connectivity between the vehicles becomes much more useful as other systems, such as autonomous driving, become more prevalent in vehicles.

Meanwhile, the onboard sensing systems will leverage improved machine learning algorithms to include several new sensors. In particular, advances in computer vision will allow the hybrid EMS to make well-informed decisions based on processing images of the road. The image can provide real-time traffic data that can be fed into the EMS, allowing it to adjust the energy strategy accordingly.

6.4 Comparison of studies

Given the critical role of HEVs and their EMS, several comprehensive reviews have examined both current and future developments in EMS technologies. For example, Saiteja et al. provided a thorough evaluation of existing rule-based and optimization-based EMSs, emphasizing the challenge of balancing model robustness with computational simplicity. Their review highlighted the increasing complexity of future EMSs, which will need to integrate data from infrastructure, geographic information systems, and other vehicles to optimize energy management. This integration will demand more sophisticated predictive models while maintaining computational efficiency. A similar study by Zhang et al. extended these insights by examining the incorporation of intelligent transportation systems into the EMS frameworks. Their analysis predicted that data from infrastructure and other vehicles would significantly enhance vehicle performance and efficiency. However, they noted that this increased data flow would also introduce considerable computational complexity, presenting new challenges for EMS optimization. Another study by Yang et al. explored conventional and future EMS approaches, focusing on the integration of vehicle-to-vehicle and vehicle-to-infrastructure connections. Their study underscored the value of DRL algorithms, which offer greater flexibility than traditional rule-based or optimization-based systems. By considering multiple parameters simultaneously, DRL allows EMSs to dynamically adapt, improving both vehicle performance and efficiency.

All of these studies point to a common trend: future EMSs will require the ability to process vast amounts of external data from vehicles, infrastructure, and geospatial sources to better predict driving conditions and optimize energy management. This is in line with the technology roadmap presented in this paper, where the focus is on leveraging additional information to achieve more comprehensive and accurate EMS predictions.

While the other reviews in the literature provide valuable insights into the evolution of EMS, this study takes a different approach. Rather than merely focusing on current EMS challenges and the benefits of future EMS systems, this study explores the natural progression of EMS as it begins to harness cutting-edge advancements in machine learning, cloud computing, computer vision, and swarm technologies. These advancements will enable future EMSs to become more adaptive, capable of real-time optimizations, and ultimately more effective in balancing performance and energy efficiency. This study not only looks at the future potential of EMS but also highlights how it will fundamentally change as new technologies continue to shape its development.

6.4.1 Research Studies

Table 6.2: Highlights of research studies

Technology Field	Impact
Machine Learning	Custom EMS algorithms that evolve with driver behaviour
Cloud Computing	Processing of large data and combining onboard and off-board sensors
Computer Vision	Predictive modelling of driving behaviour
Swarm Technology	New optimization algorithms for EMS

In the field of electric and hybrid vehicle Energy Management Systems (EMS), emerging technologies are playing a pivotal role in enhancing performance and adaptability. Machine Learning enables the development of custom EMS algorithms that continuously evolve in response to driver behaviour, ensuring personalized and efficient energy usage. Cloud Computing supports the processing of large datasets by integrating inputs from both onboard and off-board sensors, thereby enabling real-time decision-making and system updates. Computer Vision facilitates predictive modelling of driving behaviour, allowing the EMS to anticipate vehicle energy demands based on visual inputs from the surrounding environment. Additionally, Swarm Technology introduces innovative optimization algorithms that coordinate multiple parameters simultaneously, improving the overall efficiency and responsiveness of the EMS.

6.4.2 Research Advances

The evolution of the EMS for HEVs will leverage advances in a number of technology fields.

6.4.2.1 Machine Learning

Artificial intelligence (AI) algorithms allow a computer to collect and process information similar to human intelligence. Machine learning (ML) is a type of AI that allows computers to effectively learn through algorithms that evolve as more data are collected. ML is used in a variety of technology fields, including data analytics and cyber security; it has the potential to allow the EMS to improve its algorithms based on driver behaviour, traffic patterns, and weather. In particular, ML algorithms underlie DRL-based EMSs.

A research team from UC Riverside attempted to implement a DRL-based EMS into plug-in hybrid vehicles with limited success. The largest issue was that the algorithms included a substantial delay time to modify the vehicle behaviour. As such, the results from the study indicated that this EMS worked well in routes that were without variation. While ideal for public transportation, where vehicles have an identical route every day, this would cause issues for general automotive usage.

Hu et al. developed a system model for a DRL-based EMS in an HEV to autonomously learn the optimal policy based on data inputs. The study used ADVISOR, a software that models vehicle powertrains, to model the DRL-based EMS for the HEV. The researchers combined both online and offline learning techniques by training the deep-reinforced neural network offline under the urban dynamometer driving schedule and then applying the online learning under the new European driving cycle. The study found an eight percent improvement in fuel economy for the urban drive cycle and a three percent improvement for the highway drive cycle. These improvements are due to the DRL-based EMS being able to optimize the powertrain operations over the large variations seen in the urban drive cycle; meanwhile, the variations in the highway drive cycle are less substantial.

Several different research teams are evaluating ML algorithms that can further support the development of a DRL EMS. In particular, these studies seek to improve the fuel efficiency of the vehicle through an improved EMS while reducing the computation time, making the EMS more responsive. Sun et al. used a soft actor–critic scheme that sought to maximize fuel economy while still maintaining a degree of flexibility such that the algorithm was adaptable to set operating conditions. Another study by Lin et al. used an adaptive hierarchical management strategy, which is an advanced data-driven algorithm. This study found similar fuel consumption to a rule-based EMS, however, with better vehicle performance. Liu et al. and Zhou et al. used a twin delayed deep deterministic policy gradient, which optimized the EMS through penalizing irrational actions. Both studies built several simulations that found a substantial computational reduction coupled with approximately a five percent improvement in fuel consumption. Meanwhile, Tang et al. used a double DRL EMS, which sought to control the engine and transmission separately, resulting in a significant improvement in computational efficiency and a modest reduction in fuel consumption.

Using ML to support RL and DRL in EMS for HEVs offers clear advantages, but several significant challenges need to be addressed. One of the primary difficulties is the requirement for large volumes of high-quality, diverse data to train the ML models effectively. In the context of HEVs, gathering data that accurately represents the broad range of driving conditions, weather variations, and vehicle states is a complex task. Incomplete or biased data can lead to models that underperform or fail to generalize to real-world scenarios. Moreover, creating comprehensive datasets can be costly and time-consuming, particularly when simulating rare or extreme driving conditions necessary for training robust RL and DRL models.

Another challenge lies in the computational complexity of ML-augmented RL and DRL algorithms. These models often require significant processing power and memory, which can hinder their ability to operate in real-time within the hardware-constrained environments typical of vehicles. Ensuring that these systems remain responsive without compromising on performance or accuracy is difficult. Additionally, safety and robustness remain concerns, especially in unpredictable or edge case scenarios. RL and DRL models may behave unpredictably when confronted with conditions they have not encountered during training, which could lead to unsafe decisions or actions. Balancing exploration and exploitation, a key challenge in RL, becomes even more critical in real-world applications where mistakes can lead to costly or dangerous outcomes, requiring advanced methods to ensure system safety while maintaining optimal performance.

6.4.2.2 Cloud Computing

Cloud computing moves data into the cloud for on-demand processing, decreasing the onboard processing requirements. It is an efficient method of processing a large amount of data, especially when compiling data from numerous different sources. Cloud computing would allow for rapid processing of data for a vehicle to determine how to optimize the

EMS. Further, it also allows for the optimization of a group of vehicles as opposed to a single vehicle.

A study by Hu et al. proposed that traffic information and cloud computing in intelligent transportation systems can enhance HEV energy management since vehicles obtain real-time data through intelligent infrastructures and/or connected vehicles. The study proceeded to develop and assess a series of RL algorithms that incorporate this external data into an HEV EMS. A study by Du et al. found similar results when incorporating traffic data through cloud computing.

Several other research groups have studied the incorporation of off-board sensors for a hybrid EMS. A study by Liu et al. integrated cloud computing with RL and DRL for hybridized tracked vehicles. In their analysis, they used a parallel processing scheme, where a cloud-based EMS receives real-time data from the vehicle, which in turn generates artificial data to better train the onboard EMS. Their study found a savings in fuel economy of approximately 10 percent for their scheme when compared to conventional RL methods. Zhang et al. identified cloud computing as having the processing power necessary to determine a global optimization in real-time. Further, they found that cloud computing could allow the vehicle to optimize fuel economy and battery state of health, which is difficult to do with localized computing. Another study by Hou and Song used cloud computing to optimize battery usage and degradation. In their study, they identified that cloud computing allows for external inputs that can provide for predictive analysis.

Leveraging cloud computing to support EMS for HEVs offers significant benefits but also presents several challenges. Cloud computing can provide virtually unlimited computational power and storage, allowing EMS to process large datasets, run complex simulations, and optimize energy management strategies in real-time. This capability can enhance decision-making, leading to more efficient fuel consumption, improved battery management, and overall better vehicle performance. Furthermore, the cloud enables continuous updates and access to external data sources, such as weather forecasts and traffic patterns, allowing the EMS to dynamically adjust energy strategies based on real-world conditions. This adaptability can significantly improve energy efficiency and reduce emissions across varying driving environments.

However, there are notable difficulties in integrating cloud computing with EMS for HEVs. A primary concern is the reliance on stable, high-bandwidth communication between the vehicle and the cloud, which may not always be feasible, particularly in remote or rural areas. Latency issues can also arise, causing delays in real-time decision-making, which is critical for optimizing energy use on the go. Additionally, ensuring the security and privacy of data transmitted to and from the cloud is a major challenge, as vulnerabilities in communication channels could expose sensitive vehicle or driver information to cyberattacks. Managing data synchronization between cloud and onboard systems is another difficulty, as any misalignment between cloud-processed decisions and vehicle operations could lead to inefficiencies or unsafe driving conditions.

6.4.2.3 Computer Vision

Computer vision (CV) is a form of AI that allows computers to analyse images to understand the visual world. Many of the inputs used by drivers are related to visual cues (e.g., stop lights, heavy traffic, off-ramps). As such, computer vision would allow for the vehicle to take in similar inputs as a driver and make realistic, real-time changes to the EMS to accommodate perturbations in driving needs. Similarly, autonomous driving algorithms coupled with CV are also continuing to improve and provide more robust accommodations for changes in traffic patterns, weather, and driving needs.

Several studies have looked at incorporating computer vision with the EMS. A study from Wang et al. found an improvement in fuel economy of 4.3–8.8 percent for DRL algorithms with the inclusion of visual information based on simulated highway and urban drive cycles. Their study included traffic as well as stop lights that were assessed through visual means. A schematic of their study is shown in fig. Other studies on computer vision have found similar improvements in fuel economy.

Research by Chen et al. identified that a 10 percent reduction in fuel economy is possible, with no processing lag, by using cameras to identify road types and conditions. Another study by Zhang et al. uses computer vision to predict future vehicle propulsion needs, including potential traffic issues, and provides that data to the EMS. A research study by Tang et al. developed models to analyse the use of computer vision with DRL to support both car following in addition to feeding traffic data into the EMS.

Using computer vision to support EMS for HEVs offers significant advantages but also presents unique challenges. Computer vision enables the EMS to analyse visual data from onboard cameras and sensors, providing real-time insights into the vehicle's surroundings. This capability allows the EMS to make more informed decisions by detecting road conditions, traffic patterns, and obstacles, helping optimize energy usage based on the driving environment. For instance, by recognizing stop-and-go traffic or upcoming terrain changes, the system can adjust power distribution to the electric motor and internal combustion engine, improving fuel efficiency and battery life. Computer vision also supports advanced driver-assistance systems, contributing to more efficient energy management by enabling smoother driving patterns and better control over acceleration and braking.

Despite these benefits, integrating computer vision into EMS for HEVs poses several challenges. The first challenge is the computational demand required to process high-resolution images and video streams in real-time, especially in resource-constrained vehicle environments. Computer vision algorithms, particularly those based on deep learning, require significant processing power, which can strain the vehicle's hardware and limit real-time responsiveness. Additionally, ensuring the reliability of computer vision systems in diverse and unpredictable conditions, such as low light, heavy rain, or fog, is a critical concern. Vision-based EMS must be able to handle these variations while maintaining accurate environmental perception, which is difficult to achieve consistently. Another challenge is the potential for false positives or misinterpretation of visual data, leading to incorrect decisions that could negatively impact energy efficiency or vehicle safety.

Swarm Technology-Another technological field that will play a role in the long term is swarm technology. Swarm technology allows for multiple systems to work together. Although swarm technology is primarily aimed towards drones, some of the optimization techniques can be applied to EMS. In particular, swarm technology is tied to vehicle-to-vehicle communication, where the EMS can optimize globally across a group of vehicles. Zhang et al. highlighted how swarm-based vehicle-to-vehicle (V2V) communication can optimize future EMSs for HEVs by allowing vehicles to share real-time data, such as speed and road conditions, to improve energy management. Using a model predictive control framework, the study optimized torque split and gearshift in parallel HEVs, enhancing fuel economy while minimizing gearshift frequency. By simplifying the gearshift assumption over prediction horizons, the system achieves computational efficiency close to traditional methods. The study found that this integration of V2V communication into swarm-based EMS frameworks shows potential for significant improvements in real-time energy optimization and vehicle performance. Swarm technology can greatly enhance EMS for HEVs by enabling vehicles to communicate and collaborate with each other, optimizing energy use across a network of vehicles. This collective decision-making allows for more efficient route planning, traffic management, and energy distribution, as vehicles can share information about road conditions, traffic patterns, and available charging infrastructure. By functioning as part of a coordinated system, HEVs can reduce energy consumption and improve battery performance, particularly in congested urban environments where real-time coordination is essential.

However, implementing swarm technology in EMS also presents challenges. One of the primary difficulties is maintaining reliable, low-latency communication between vehicles, especially in high-density traffic or remote areas. Ensuring data synchronization across multiple vehicles in dynamic environments is also complex, as delays or inconsistencies could lead to inefficient or conflicting decisions. Additionally, securing the network of interconnected vehicles against cyber threats is critical, as a breach in communication protocols could disrupt the entire swarm, leading to potential safety hazards or energy inefficiencies. The integration of ML, cloud computing, computer vision, and swarm technology will significantly advance the evolution of EMS for HEVs. ML and cloud computing will play a crucial role in establishing and enhancing RL and DRL schemes within EMS. ML allows for adaptive learning based on real-time and historical data, optimizing energy usage and powertrain control to improve fuel efficiency and performance. Cloud computing, with its vast computational resources, enables the continuous refinement of these algorithms, allowing the EMS to adapt dynamically to changing driving conditions by leveraging real-time data from multiple sources. Together, ML and cloud computing will facilitate more responsive and intelligent EMS systems capable of optimizing energy management across diverse driving scenarios. The addition of computer vision and swarm technology into EMS will further expand the system's capabilities. Computer vision will enable vehicles to better understand their surroundings by processing visual data from onboard sensors, allowing the EMS to adjust energy strategies based on road conditions, traffic, and obstacles. Swarm technology will support inter-vehicle communication, enabling coordinated energy management across a fleet of HEVs. By sharing data on road conditions, traffic patterns, and charging infrastructure, the EMS in each vehicle can optimize its performance within a collective network, reducing overall energy consumption. While implementing these technologies presents challenges, substantial progress in other sectors, such as autonomous driving and networked systems, is paving the way for their integration into HEV EMS, providing new opportunities for efficiency and performance improvements.

Hybrid electric vehicles (HEVs) are set to play a critical role in the future of the automotive industry. To operate efficiently, HEVs require a robust energy management strategy (EMS) that decides whether the vehicle is powered by the engine or electric motors while managing the battery's state of charge. The EMS must rapidly adapt to driver demands and optimize energy usage, ideally predicting battery charge rates and fuel consumption to adjust the powertrain in real time, even under unpredictable driving conditions. As HEVs become more prevalent, EMS technologies will advance to improve predictive capabilities. This analysis provides an overview of current EMS systems, including both rule-based and optimization-based approaches. It explores the evolution of EMS development through a technology roadmap, highlighting the integration of advanced algorithms such as reinforcement learning and deep learning. The analysis addresses the technologies that underly this evolution, including machine learning, cloud computing, computer vision, and swarm technology. Key advances and challenges in these technologies are discussed, along with their implications for the next generation of EMS systems for HEVs. The analysis of these technologies indicates that they will play a key role in the evolution of EMS technology, allowing it to better optimize driver needs and fuel economy.

The automotive industry has been powered by the internal combustion engine for over a century. However, the industry is moving away from traditional internal combustion engines due to concerns about climate change and foreign oil dependence. Recently, many countries have banned the sale of automobiles that use only internal combustion engines due to harmful emissions. While the end goal is for the entire automotive sector to transition to electric vehicles (EVs) powered by renewable energy, this shift is currently limited by the cost of batteries and the associated vehicle range. Meanwhile, hybrid electric vehicles (HEVs) can achieve many of the advantages of EVs while avoiding issues related to range. HEVs use both an internal combustion engine and a battery/motor assembly to power the vehicle.

Inherent in HEVs is their energy management strategy (EMS), which sets the conditions for when the vehicle is powered by the engine versus the battery/motor assembly. The EMS plays a key role in ensuring the vehicle operates under optimal conditions to minimize fuel consumption. The goal of this review paper is to provide a technology roadmap for the

EMSs used in HEVs. It first discusses the significance and current state of the EMSs in modern HEVs. It then develops a technology roadmap for future EMSs based on market trends. This paper then presents a comprehensive review of the technologies driving the evolution of EMSs, including machine learning, computer vision, cloud computing, and swarm technology. These technologies are discussed in relation to EMSs, with a particular focus on how their advances affect overall fuel and energy consumption.

6.5 Overview of Energy Management Systems

6.5.1 Importance of Hybrid Technologies

In theory, EVs offer better performance, lower operating costs, and help achieve carbon neutrality when compared to traditional vehicles. However, they also come with significant challenges. One of the biggest issues is the size and cost of the requisite battery pack. Currently, these packs use Li-ion batteries with an energy density of about 220 Whr/kg. For example, the Tesla Model S has a 120-kWh battery pack weighing roughly 540 kg. For USD 118 per kWh, this results in a USD 14,160 battery pack for the Model S. The high cost of the battery pack makes EVs prohibitively expensive for many consumers.

Another issue is the weight of the battery pack, which affects the vehicle's overall energy consumption since the power required for locomotion scales with vehicle weight. For example, a Toyota Camry weighs 1545 kg, whereas a Tesla Model S weighs about 2045 kg. This 32% weight increase leads to an estimated 25% rise in the amount of power required for locomotion. In traditional vehicles, energy is generated onboard by the engine. In EVs, energy is produced off-site and stored in the battery, but this introduces inefficiencies, such as transmission and charging losses, each about 90% efficient. Therefore, a Tesla Model S requires 54% more energy from the grid than what the Toyota Camry would need from its onboard engine. Although grid energy is generally cleaner and more efficient, these extra demands can offset some of the environmental benefits of EVs.

HEVs mitigate the cost and weight issues by using smaller battery packs and downsized engines. For instance, the Toyota Camry Hybrid weighs only 45 kg more than the standard Camry since it only has a 1 kWh battery pack. The minimal weight difference means that the power required by the HEV for locomotion is comparable to the traditional version. Additionally, the HEV benefits from regenerative braking and uses battery power when the engine is least efficient. As a result, the fuel economy improves from 28 mpg city/39 mpg highway to 51 mpg city/53 mpg highway. This increase in fuel efficiency contributes significantly to carbon neutrality goals by 2050.

Assuming a mix of city and highway driving and USD 4 per gallon of gasoline, the improved fuel economy of the hybrid Camry translates to savings of around USD 4600 over 100,000 miles compared to the non-hybrid variant of the Camry. Although these savings may not fully offset the hybrid's higher initial cost at present, rising fuel prices, government incentives, and decreasing battery costs are expected to make the total cost of ownership for HEVs lower than that of their non-electrified counterparts in the near future. Moreover, the increased fuel economy of the HEV results in a significant reduction in fossil fuels used and carbon emitted when compared to a traditional vehicle.

6.5.2 Hybrid Topologies

As shown in fig, HEVs typically use parallel, series, or power-split architectures based on the engine and battery/motor configuration. Even traditional vehicles feature a degree of hybridization, as the battery/motor starts the engine, though it doesn't assist with locomotion.

6.5.3 Classification of EMS in Electric and Hybrid Vehicles GTUexam-3170923 WIN2025

Energy Management Strategies (EMS) are control approaches designed to optimize the power flow between various energy sources and the load in Electric Vehicles (EVs) and Hybrid Electric Vehicles (HEVs). The main objectives are to maximize efficiency, prolong battery life, and minimize fuel consumption and emissions.

1. **Rule-Based Strategies-** These strategies rely on predefined control rules or lookup tables based on driving conditions. a) Deterministic Rule-Based EMS: Uses fixed logic, such as 'Use battery at low speeds and ICE at high speeds'. Example: State machine control. b) Fuzzy Logic-Based EMS: Handles uncertainties with fuzzy rules, allowing smooth transitions between power sources.

2. **Optimization-Based Strategies-** These use mathematical optimization to minimize fuel consumption, emissions, or energy losses. Global Optimization: Uses full trip information (e.g., Dynamic Programming). Highly accurate but computationally intensive) Real-Time Optimization: Approximation of global methods for real-time applications (e.g., Equivalent Consumption Minimization Strategy – ECMS).

3. **Learning-Based Strategies-** Machine learning and AI are applied to dynamically optimize EMS for varying driving conditions. A) Neural Network-Based EMS: Predicts optimal energy split using trained models. Reinforcement Learning (RL)-Based EMS: Learns optimal control policies by interacting with the environment.

4. **Hybrid Strategies-** Combines rule-based, optimization-based, and learning-based methods to balance accuracy, computational cost, and adaptability.

5. **Technology** In light of the challenges outlined in the preceding section, Energy Management Systems (EMSs) are expected to advance toward more effective optimization of fuel consumption and overall vehicle performance. The figure illustrates a proposed evolutionary roadmap for EMSs in hybrid electric vehicles (HEVs), reflecting ongoing progress in enabling technologies. Specifically, existing rule-based and optimization-based EMS approaches are anticipated to be superseded by more sophisticated systems capable of adapting to individual driving behaviours. In the longer term,

EMS development is expected to incorporate a wide range of external influences, including traffic and weather. Each of these stages of EMS technology will be discussed in the subsequent. sections.

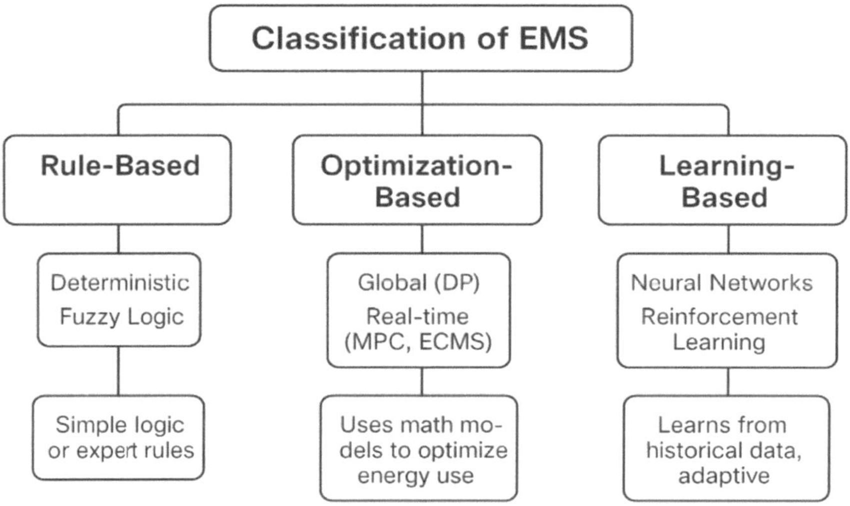

Fig.6.6: Classification of EMS

Finally, the Learning-Based strategies incorporate modern Artificial Intelligence techniques such as Neural Networks and Reinforcement Learning. These methods have the ability to learn from historical driving data and adapt to changing operating conditions, making them highly flexible and intelligent in decision-making. Their adaptability makes them promising for future vehicle applications, though they require large datasets and significant training time.

Within this roadmap, the term long-term is broadly interpreted as a horizon of approximately 20 years. By the conclusion of this timeframe, hybrid electric vehicles (HEVs) could be displaced by fully electric vehicles enabled by advanced battery chemistries. Such next-generation batteries are expected to mitigate the weight and cost limitations inherent to current lithium-ion technologies. Moreover, projecting technological developments beyond a 20-year horizon in a rapidly evolving domain involves considerable uncertainty. Energy Management Strategies (EMS) in Electric and Hybrid Electric Vehicles can broadly be classified into three main categories: Rule-Based, Optimization-Based, and Learning-Based approaches. The Rule-Based strategies rely on either deterministic rules or fuzzy logic. These methods use simple logic or expert-defined rules to manage the flow of energy between the power sources and the load. Their major advantage lies in their ease of implementation and fast decision-making, though they may lack adaptability to complex driving conditions. The Optimization-Based strategies are more mathematically intensive and aim to achieve either global or real-time optimization of energy use. Global optimization methods, such as Dynamic Programming (DP), provide benchmark solutions but are computationally heavy and often offline. On the other hand, real-time optimization approaches like Model Predictive Control (MPC) and Equivalent Consumption Minimization Strategy (ECMS) strike a balance by providing practical, online solutions that enhance fuel efficiency and reduce emissions.

6.5.3.1 Near-Term Energy Management Systems

Existing rule-based and optimization-based Energy Management Systems (EMSs) generally do not incorporate variations in individual driving behavior. In the near future, however, EMSs are expected to adapt more effectively to distinct driving patterns and operating conditions, as illustrated in Figure. Among emerging approaches, reinforcement learning (RL)-based EMSs represent a promising method for advanced energy management. This technique leverages historical driving data to learn control strategies and subsequently applies the acquired knowledge during vehicle operation. An RL-based EMS can be designed to compute an optimal control policy by utilizing inputs such as vehicle speed, driver power demand, and battery state of charge (SOC) to regulate engine power. By accommodating uncertain and time-varying driving cycles, this adaptive approach can achieve reduced fuel consumption when compared to traditional dynamic programming methods.

At a fundamental level, a reinforcement learning (RL) framework is composed of two primary components: an agent and an environment. The agent repeatedly interacts with the environment by observing system inputs and selecting corresponding actions. Each selected action causes the environment to transition to a new state, after which a reward associated with that state transition is computed and returned to the agent. The objective of RL is to enable the agent to maximize its cumulative reward by learning an effective interaction policy. This interaction and learning process occurs iteratively over time. Despite its advantages, RL presents several challenges. An excessively large state space can lead to computational complexity and reduced learning efficiency. Moreover, RL methods are often described as *data-intensive*, as they require substantial amounts of data and continuous computation. Learning is typically driven by a scalar reward signal, which may be sparse, delayed, or noisy, further complicating the training process.

An emerging approach to address these limitations involves integrating RL with deep learning (DL), resulting in deep reinforcement learning (DRL). Although both DL and RL fall under the broader category of machine learning, they differ fundamentally in methodology. Deep learning relies on training datasets to extract features and generalize knowledge to unseen data, whereas reinforcement learning emphasizes learning optimal behavior through interaction and trial-and-error exploration. DL techniques enable the identification of complex patterns by analyzing current data and automatically extracting salient features. When combined, DRL mitigates the issue of state-space explosion while maintaining the flexibility required for dynamic and. Consequently, deep reinforcement learning (DRL) has gained widespread adoption in domains characterized by uncertain and dynamic environments, including robotics, heating, ventilation, and air-conditioning (HVAC) control, traffic ramp metering, and other complex control applications. In the automotive field, DRL has been successfully applied to advanced driver-assistance systems, such as lane-keeping assistance and autonomous braking functions.

6.5.3.2 Mid/Long-Term EMS

Figure displays the state diagram for the mid/long-term EMS for HEV. The major change from the near-term systems is that in the long term, the EMS will be offloaded from the vehicle through the use of cloud computing. In doing so, the system can take multiple complex inputs from external sources, such as traffic and weather. With the processing partially offloaded, the EMS can then fuse all of this data with onboard vehicle data to account for individual driving behaviours.

In addition, the Energy Management System (EMS) can utilize information exchanged with nearby vehicles, particularly those traveling ahead of the host vehicle. Through such connectivity, the EMS can access information that may not be obtainable solely from onboard sensors or conventional traffic data sources. For instance, the system can recognize the precise moment when a leading vehicle initiates braking. Furthermore, it can identify unforeseen external loads acting on the vehicle, such as sudden wind gusts or variations in road gradient. As vehicle connectivity continues to expand—especially alongside the growing adoption of autonomous driving technologies—the value and effectiveness of inter-vehicle data sharing for EMS decision-making will increase substantially.

At the same time, onboard sensing capabilities are expected to benefit from advances in machine learning, enabling the integration of additional sensor modalities. In particular, progress in computer vision techniques will allow hybrid EMSs to make more informed decisions through real-time analysis of road imagery. Visual data can provide immediate insights into traffic conditions, which can be incorporated into the EMS to dynamically adjust energy management strategies.

6.5.4 Comparison with Related Studies

Given the importance of hybrid electric vehicles (HEVs) and their associated EMSs, numerous comprehensive studies have investigated both existing and emerging energy management technologies. For example, Saiteja *et al.* conducted an extensive review of rule-based and optimization-based EMS approaches, highlighting the inherent trade-off between model robustness and computational simplicity. Their work emphasized that future EMSs are likely to become increasingly complex, requiring the integration of data from infrastructure systems, geographic information systems, and neighbouring vehicles to achieve improved energy optimization. Such integration necessitates more advanced predictive models while still maintaining computational efficiency.

Similarly, Zhang *et al.* expanded upon these findings by examining the incorporation of intelligent transportation systems within EMS frameworks. Their study suggested that information obtained from infrastructure and surrounding vehicles could significantly enhance vehicle efficiency and performance. However, they also noted that the growing volume of data would introduce substantial computational demands, thereby posing new challenges for EMS design and optimization.

In another related investigation, Yang *et al.* analysed both conventional and next-generation EMS strategies, with particular emphasis on vehicle-to-vehicle and vehicle-to-infrastructure communication. Their results highlighted the advantages of deep reinforcement learning (DRL) algorithms, which provide greater adaptability than traditional rule-based or optimization-based methods. By simultaneously considering multiple system variables, DRL-enabled EMSs can dynamically respond to changing conditions, leading to improvements in overall vehicle efficiency and performance.

Although existing literature reviews offer important perspectives on the development of Energy Management Systems (EMS), this study adopts a distinct viewpoint. Instead of concentrating solely on present-day EMS challenges or anticipated benefits of future systems, it examines the organic evolution of EMS as it increasingly integrates advanced technologies such as machine learning, cloud computing, computer vision, and swarm intelligence. The incorporation of these innovations is expected to make future EMS more adaptive, enable real-time optimization, and enhance their ability to balance system performance with energy efficiency. Beyond discussing future possibilities, this study emphasizes how the ongoing integration of emerging technologies will fundamentally transform the design and operation of EMS.

6.6 Operating Modes of a Parallel Hybrid Electric Vehicle

A parallel hybrid electric vehicle operates in four major modes: Recharging, Boosting, Recuperation, and Electric Driving. A parallel hybrid electric vehicle can operate in multiple modes depending on driving conditions and energy requirements. In the recharging mode, the internal combustion engine not only drives the vehicle but also powers the electric motor, which works as a generator to recharge the battery. This ensures that the battery remains adequately charged during operation. The boosting mode occurs when both the engine and the electric motor supply torque to the wheels simultaneously, providing additional power during situations such as acceleration, overtaking, or hill climbing. Another important mode is recuperation, in which the kinetic energy of the wheels during braking or deceleration is captured by the electric motor, which functions as a generator and stores the recovered energy in the battery. Finally, in the electric driving mode, the vehicle operates solely on the electric motor powered by the battery, while the engine remains switched off. This mode is typically used for low-speed or urban driving, allowing for quiet and zero-emission operation.

Recharging - In the Recharging mode, the internal combustion engine drives the vehicle and simultaneously powers the electric motor, which functions as a generator to recharge the battery.

Boosting- In the *Boosting mode*, both the engine and the electric motor supply torque to the wheels at the same time, thereby providing additional power during acceleration, overtaking, or hill climbing.

Recuperation- The *Recuperation mode* occurs during braking or deceleration, when the kinetic energy of the wheels is captured by the electric motor. Acting as a generator, the motor converts this energy into electricity and stores it in the battery.

Finally, in the *Electric Driving mode*, the vehicle runs solely on the electric motor powered by the battery, while the engine remains switched off. This mode is typically used for low-speed or urban driving, enabling quiet and zero-emission operation. These operating modes make hybrid vehicles highly flexible, energy-efficient, and well-suited for future sustainable transportation

6.6.1 Implementation Issues

The practical implementation of Energy Management Strategies (EMS) in Electric and Hybrid Electric Vehicles encounters several challenges. One of the primary concerns is computational complexity, as real-time decision-making requires fast and efficient processors capable of handling large datasets and executing control algorithms within milliseconds. Another issue arises from hardware constraints, such as limited sensor resolution and actuator delays, which can reduce the accuracy and responsiveness of the system. Furthermore, uncertainty handling becomes critical since driving conditions, ambient temperature, and component aging introduce unpredictable variations that must be managed effectively to ensure optimal performance. Lastly, driver behaviour plays a significant role, as the EMS must adapt dynamically to different driving styles, ranging from aggressive acceleration to eco-friendly driving patterns, in order to balance efficiency, performance, and comfort.

6.6.2 Charging Techniques for Electric Vehicles

Electric Vehicles (EVs) employ different charging techniques depending on the required speed, infrastructure, and application. Level 1 charging, based on a 120 V single-phase AC supply, is primarily used in residential settings. It is the slowest method, often requiring 8 to 20 hours to fully charge the battery, making it suitable for overnight charging at home. Level 2 charging operates on a 240 V single-phase or three-phase AC supply and is widely adopted in public charging stations. With a shorter charging time of 4 to 8 hours, it offers a more practical solution for daily commuting needs. Level 3 charging, also known as DC fast charging, functions at a much higher voltage range of 400 V to 900 V DC. It is designed for commercial fast-charging applications, where vehicles can be charged within 30 to 60 minutes, making it highly suitable for long-distance travel and fleet operations. In addition to these wired methods, wireless charging (inductive charging) has emerged as an innovative approach, where energy is transferred through electromagnetic fields. This technique is being explored in smart parking systems and limited-use cases, with the charging time largely dependent on the specific technology employed.

Table 6.3: Charging techniques

Type of Charging	Voltage Level	Application	Time Required
Level 1 (AC)	120V (1-phase)	Home charging	8–20 hrs
Level 2 (AC)	240V (1/3-phase)	Public charging stations	4–8 hrs
Level 3 (DC Fast)	400V–900V (DC)	Commercial fast-charging	30–60 mins
Wireless (Inductive)	Varies	Smart parking, limited use	Depends on tech

The charging process of electric vehicles (EVs) is strongly dependent on the charging level and the power rating of the charger. The relationship can be visualized using a graph between charging time (hours) on the vertical axis and charging power (kW) on the horizontal axis. Level 1 Charging (AC slow charging) operates at the lowest power range, typically around 1–2 kW. Since the charging rate is low, the time required to fully charge the battery is the longest, often taking 8–12 hours. This method is usually employed with household outlets and is suitable for overnight charging. Level 2 Charging (AC fast charging) provides a higher power output, usually between 3–20 kW, significantly reducing charging time to about 3–6 hours. It is commonly used at residential charging stations and workplace charging points. Level 3 Charging (DC fast charging) offers the highest power levels, typically 50 kW and above, allowing the battery to charge up to 80% in less than 1 hour. This method is preferred for long-distance travel where rapid recharging is necessary.

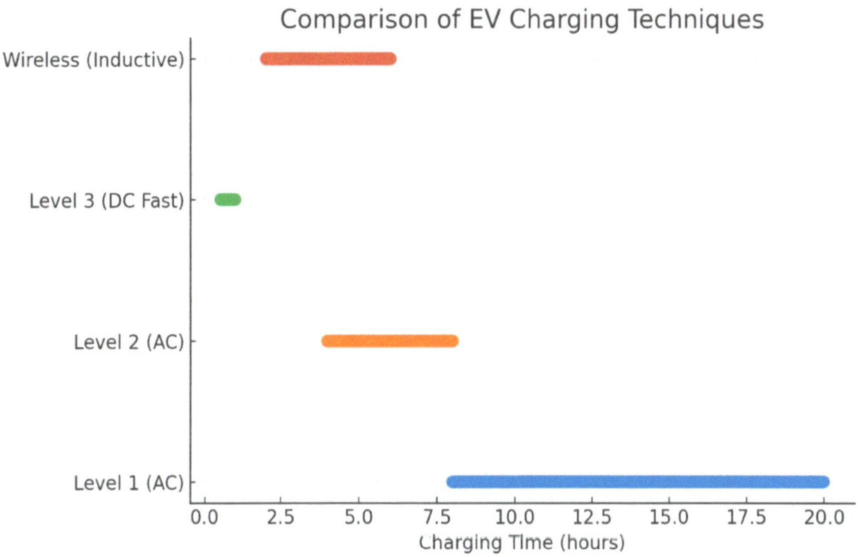

Fig.6.7: Comparison of EV charging techniques

6.7 Charging Station Layout

A modern electric vehicle (EV) charging station is composed of several integrated components that ensure reliable, efficient, and safe operation. The power supply is typically derived either from the utility grid or renewable energy sources such as solar panels or wind turbines, often connected through a transformer for proper voltage matching. An optional energy storage system, consisting of battery banks or supercapacitors, may be included to store excess energy, support peak shaving, and provide backup power when required. The power conversion units play a vital role in conditioning the energy flow: an AC/DC rectifier converts the alternating current from the grid into direct current, a DC/DC converter regulates the voltage and current to match the EV battery's requirements, and in vehicle-to-grid (V2G) systems, an inverter converts DC back into AC to send power back to the grid. The charging points and connectors provide the interface with the vehicle, offering different charging levels such as Level 1, Level 2 (AC), and Level 3 DC fast charging, and are equipped with standardized connectors like CCS, CHAdeMO, and Type 2. A dedicated communication and control unit ensures seamless interaction between the vehicle's battery management system (BMS) and the charging station, while also handling authentication, payment integration, and smart grid connectivity. To guarantee safety, protection devices including circuit breakers, fuses, surge protectors, and earthing arrangements are incorporated, safeguarding both users and equipment against electrical faults and hazards.

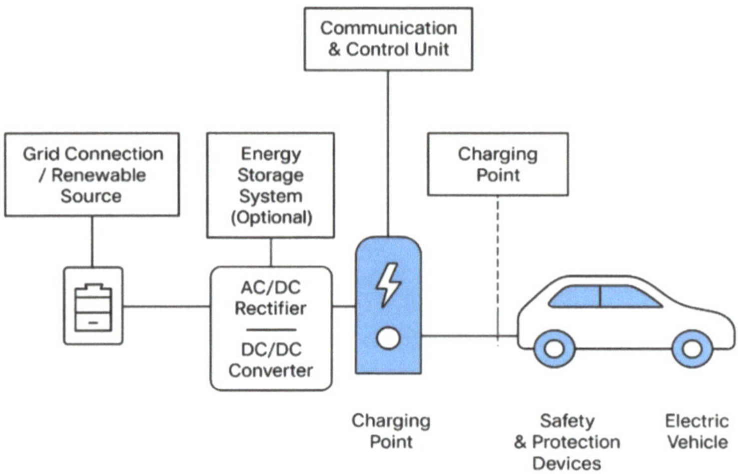

Fig.6.8: Charging station

Safety and protection devices such as circuit breakers, fuses, surge protectors, and earthing arrangements play a crucial role in ensuring both user safety and the protection of equipment. These components safeguard the charging system from electrical faults, overloads, short circuits, and voltage surges, thereby preventing potential hazards and maintaining reliable operation of the charging station.

Example: EMS in Toyota Prius

The Toyota Prius is a classic example of how Energy Management Strategies (EMS) are implemented in hybrid vehicles. It primarily employs a rule-based EMS, where the control logic follows pre-defined operating conditions. At low vehicle speeds, the system prioritizes the use of the electric motor to ensure quiet and efficient operation while minimizing fuel consumption. As the speed increases or higher power demand arises, the control strategy shifts to the internal combustion engine (ICE), either independently or in combination with the electric motor, to provide the necessary performance.

Additionally, the Prius effectively utilizes regenerative braking, where the kinetic energy during braking is converted into electrical energy and stored back in the battery. This integration of multiple operating modes highlights the efficiency of rule-based EMS in balancing fuel economy, performance, and sustainability.

- Uses rule-based EMS
- Prioritizes electric mode at low speed
- Switches to ICE at higher speeds
- Regenerative braking to charge the battery

Summary

Concept	***Key***
EMS Role	*Manages power sources for efficiency*
Types of EMS	*Rule-based, Optimization-based, Learning-based*
Implementation Challenges	*Complexity, sensor limits, driver variability*
Charging Techniques	*Level 1/2/3, Inductive charging*

Energy Management Strategies (EMS) play a crucial role in Electric and Hybrid Electric Vehicles by managing power flow among different energy sources to maximize efficiency, performance, and sustainability. Different types of EMS have been developed, including rule-based strategies, which follow predefined operating conditions, optimization-based strategies, which rely on mathematical models to achieve global or real-time efficiency, and learning-based strategies, which adapt over time using data-driven approaches. Despite their benefits, the implementation of EMS faces several challenges, such as computational complexity, hardware and sensor limitations, and variability in driver behaviour. Alongside EMS, charging techniques for EVs also play a vital role in their practicality. These range from Level 1 (home charging) and Level 2 (public charging stations) to Level 3 DC fast charging for quick replenishment, while inductive wireless charging is emerging as an innovative alternative for future mobility.

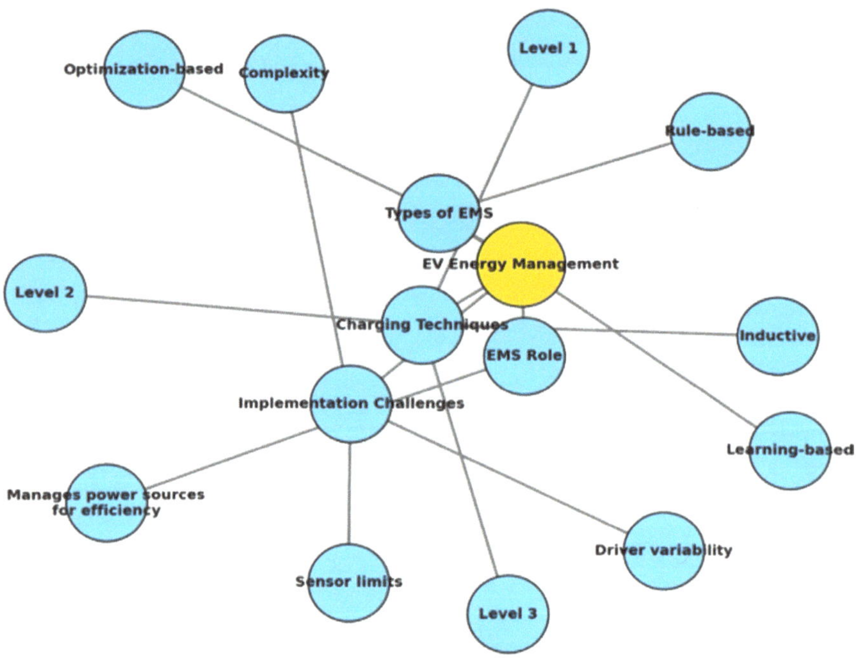

Fig.6.9: Mapping diagram of energy management in EV

Remembrance-*ENERGY, EMISSION, BATTERY, CLASSIFICATION OF EMF*

Keywords: *energy management systems; technology roadmap; deep learning; reinforcement learning; hybrid electric vehicles*

Multiple Choice Questions

1. **What is the primary objective of Energy Management Strategies (EMS) in hybrid and electric vehicles?**
   a) To increase vehicle weight
   b) To optimize power flow between energy sources and load ✓
   c) To reduce battery voltage rating
   d) To improve vehicle aerodynamics
2. **Which of the following is a characteristic of Rule-Based EMS?**
   a) Uses neural networks for adaptive learning
   b) Relies on expert-defined rules and fuzzy logic ✓
   c) Requires large datasets for training
   d) Implements dynamic programming for global optimization
3. **Dynamic Programming (DP) is commonly used in which category of EMS?**
   a) Rule-Based
   b) Optimization-Based ✓
   c) Learning-Based
   d) Hybrid EMS
4. **Which of the following is NOT a limitation of Optimization-Based EMS in real-time use?**
   a) High computational demand
   b) Requirement of predictive models
   c) Limited adaptability
   d) Inability to provide benchmark solutions ✓
5. **Learning-Based EMS, such as those using reinforcement learning, mainly depend on:**
   a) Real-time rule generation
   b) Predefined mathematical models
   c) Historical data and adaptive learning ✓
   d) Fixed duty cycles

6. In the comparison of EMS types, which strategy is easiest to implement but least adaptive?

a) Rule-Based ☑

b) Optimization-Based

c) Learning-Based

d) Multi-Objective EMS

7. Which of the following is an implementation challenge for EMS in hybrid vehicles?

a) High fuel efficiency

b) SOC estimation accuracy ☑

c) Low emissions

d) Lightweight components

8. Level 3 charging (DC fast charging) typically delivers power in the range of:

a) 1–2 kW

b) 3–20 kW

c) 50 kW and above ☑

d) 0.5 kW

9. Which component in a charging station converts AC grid supply into DC?

a) Inverter

b) DC/DC Converter

c) AC/DC Rectifier ☑

d) Transformer

10. The Communication and Control Unit in an EV charging station is primarily responsible for:

a) Voltage step-up

b) Thermal management

c) Interaction with the Battery Management System (BMS) ☑

d) Increasing charging current

11. Which safety device is used in charging stations to protect against sudden voltage spikes?

a) Circuit breaker
b) Fuse
c) Surge protector ✓
d) Earthing rod

12. Vehicle-to-Grid (V2G) functionality in charging stations requires which power conversion unit?

a) Rectifier
b) Inverter ✓
c) Transformer
d) Converter-less system

India-GTU Questions

List the optimization-based strategies in Hybrid Electrical Vehicles.	chapt6pg182 SUM2025	**04**
Explain the Energy Management concept for HEV.	chapt6pg177	**03**
List the optimization-based strategies in Hybrid Electrical Vehicles?	chapt6pg182 WIN2024	**04**
Give a classification of different energy management strategies.	chapt6pg201,203 WIN2024	**04**

Chapter 07

Hybridation-Future Scope & Smart Energy Integration

7 HYBRIDATION-FUTURE SCOPE AND SMART ENERGY INTEGRATION

7.1 Different Energy Integrations

Vehicle-to-Grid (V2G) is a technology that enables an electric vehicle not only to draw energy from the grid but also to send electricity back, helping in load balancing, supporting renewable energy integration, and providing backup during peak demand through a bidirectional charger. Vehicle-to-Load (V2L) allows an EV to act as a mobile power source by directly supplying electricity to external appliances or devices, making it useful for camping, outdoor work, or emergency relief situations. Vehicle-to-Vehicle (V2V), on the other hand, enables one EV to transfer energy to another using bidirectional power converters, providing emergency roadside charging or range extension when charging stations are unavailable. Together, V2G, V2L, and V2V enhance the utility, resilience, and flexibility of EVs beyond conventional Vehicle to Grid (V2G).

7.1.1 Vehicle to Grid (V2G)

Vehicle-to-Grid (V2G) systems enable a bidirectional flow of energy between an electric vehicle and the power grid using the vehicle's battery as a distributed energy source. In this setup, the stored battery energy of an EV can be discharged back to the grid during periods of peak demand, supporting electricity supply and improving grid stability. V2G offers multiple benefits: it provides grid support through services such as frequency regulation and load balancing, ensures better integration of renewable energy by storing and releasing energy when required, and offers financial advantages to EV owners in the form of reduced electricity costs or incentives for participating in V2G programs. Thus, V2G not only enhances energy security but also strengthens the role of EVs in sustainable power systems.

7.1.2 Vehicle to Load (V2L)

V2L technology allows the electricity stored in an EV's battery to directly power other equipment, appliances, or devices. An onboard inverter converts the DC (direct current) power from the EV's battery into AC (alternating current) power, which is compatible with most standard devices. It is used in Camping and Off-Grid Power for Powering lights, heaters, cooking appliances, and other devices at remote locations. It is used in Backup Power for providing electricity to a home or building during a power outage also used in tool Powering for supplying power to tools and equipment for construction or other work.

7.1.3 Vehicle to Vehicle (V2V)

Vehicle-to-Vehicle (V2V) is a system that enables direct communication and power transfer between multiple vehicles. Through this technology, EVs can broadcast safety-related information such as sudden braking, road hazards, or collision warnings to nearby vehicles, thereby improving traffic safety and reducing accident risks. In addition to communication, V2V also supports energy transfer, where one EV can provide emergency charging to another that is running low on battery, offering a supplementary power source in critical situations.

Thus, V2V enhances both safety on the road and reliability of electric mobility by ensuring connectivity and mutual energy support among vehicles.

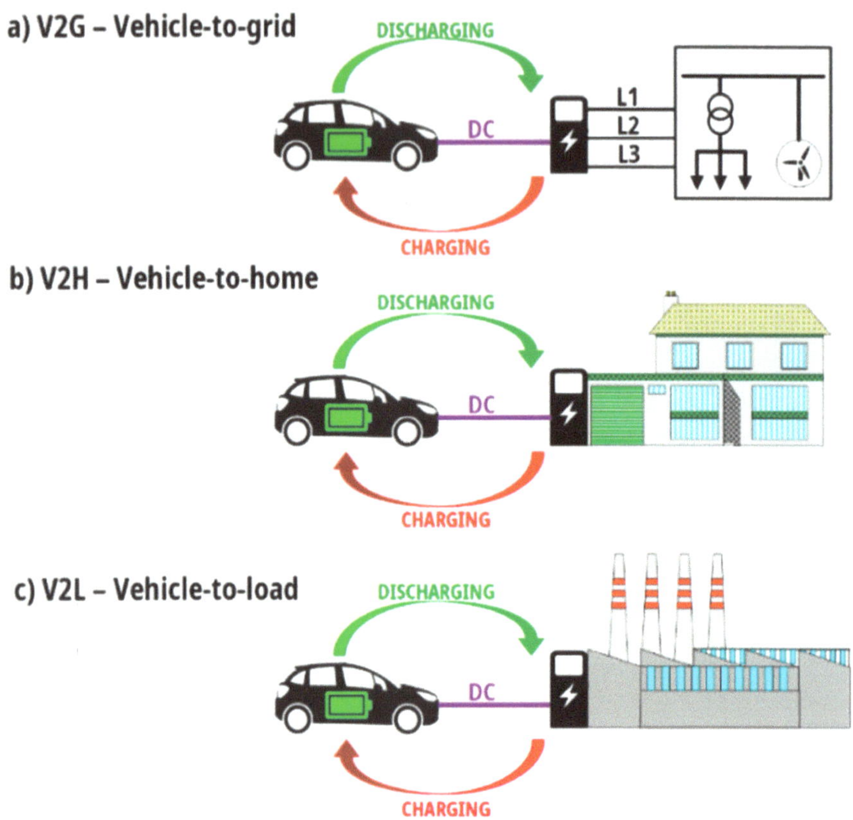

Fig.7.1: Diagram illustrating the difference between V2G,V2H,V2L

Hybridness refers to the combination of ICE & HEV GTUexam-3170923 SUM2025

Hybridness refers to the degree of combination between the internal combustion engine (ICE) and the electric drive system in a Hybrid Electric Vehicle (HEV). It indicates how much of the vehicle's total propulsion power comes from the electric motor compared to the engine.

Mathematically, it can be expressed as:

$$\text{Hybridness (H)} = \frac{P_{electric}}{P_{engine} + P_{electric}}$$

$P_{electric}$ = rated power of the electric motor

P_{engine} = rated power of the internal combustion engine

$H \approx 0 \rightarrow$ Vehicle is almost like a conventional ICE vehicle.

$H \approx 1H \rightarrow$ Vehicle behaves like a Battery Electric Vehicle (BEV).

$0<H<10 < H < 10<H<1 \rightarrow$ Degree of hybridization (mild, full, or plug-in HEV).

Example:

If an HEV has Engine power = 60 kW & Motor power = 40 kW, what is the hybridness?

$$H = \frac{40}{60 + 40} = 0.4$$

This means the vehicle has 40% hybridness, i.e., 40% contribution from the electric drive.

7.2 Charging Infrastructure and Smart Grid Integration

This topic relates to bidirectional Power Flow & Smart Energy Integration

- These terms relate to how EVs interact with external systems for power supply, energy storage, or data transfer.
- **V2G** – EVs supplying power back to the grid.
- **V2L** – EVs powering external loads.
- **V2V** – Direct energy transfer between two vehicles.

Charging Infrastructure and Smart Grid Integration focuses on the intelligent connection between electric vehicles (EVs), electrical networks, and digital control systems to enable advanced energy exchange. Unlike conventional charging systems that allow only one-way power transfer, modern infrastructure supports bidirectional power flow, enabling EVs to function as both energy consumers and energy providers. This capability plays a vital role in smart energy integration, where EVs interact with smart grids and renewable energy sources to enhance energy efficiency, grid reliability, and demand-side management. A key concept within this framework is Vehicle-to-Grid (V2G), which allows EVs to feed stored electrical energy back into the grid during peak demand periods, helping balance supply and load. Vehicle-to-Load (V2L) enables EVs to supply power directly to external devices or emergency loads, improving energy flexibility. Additionally, Vehicle-to-Vehicle (V2V) technology supports direct power transfer between electric vehicles, useful for energy sharing and roadside assistance. These functionalities are enabled by advanced power electronics, communication interfaces, and smart control strategies. Overall, integrating charging infrastructure with smart grids transforms EVs into active energy assets, supporting sustainable power systems and the efficient use of electrical energy.

7.3 Early version of EHV

A Hybrid Electric Vehicle (HEV) is a type of automobile that utilizes both an internal combustion engine (ICE) and an electric motor powered by a battery to operate. This dual power source enables HEVs to offer improved fuel efficiency and reduced emissions compared to traditional vehicles. HEVs blend the benefits of gasoline-powered cars with those of electric vehicles, creating a more efficient and environmentally friendly mode of transportation. The origins of electric vehicles date back to the 1830s when inventors such as Thomas Davenport and Robert Anderson developed early versions of battery-powered vehicles using non-rechargeable batteries.

Fig.7.2: Early EV

By the 1890s, electric vehicles were more popular than gasoline-powered cars, outselling them by a ratio of 10:1. Well-known automotive brands such as Oldsmobile and Studebaker began as electric vehicle companies. The first hybrid vehicle was introduced by the Krieger Company in the early 1900s. However, mass production of affordable gasoline vehicles by Ford in the 1910s led to the decline of electric vehicles, though they remained in limited use until the Great Depression.

7.4 Types of Hybridization

Hybridization in vehicles refers to the degree to which electric power is integrated with the traditional fuel system. There are three main types:

Mild (Micro) Hybridization: These vehicles feature idle-stop functions and regenerative braking but cannot use the electric motor to drive the vehicle. Their systems typically operate at 42 volts or lower.

Medium Hybridization: Along with idle-stop and regenerative braking, the electric motor assists the engine but cannot propel the vehicle on its own. These vehicles typically use batteries with voltages between 144 and 158 volts.

Full Hybridization: These vehicles can use the electric motor alone to propel the vehicle and also benefit from regenerative braking and idle-stop features. Battery voltages in full hybrids generally range from 200 to 300 volts.

Types of Hybrid Electric Vehicles (HEVs)

HEVs are categorized based on how power is delivered to the drivetrain:

Series Hybrid: In this type, the internal combustion engine powers a generator, which either charges the battery or powers the electric motor. The engine does not directly drive the wheels.

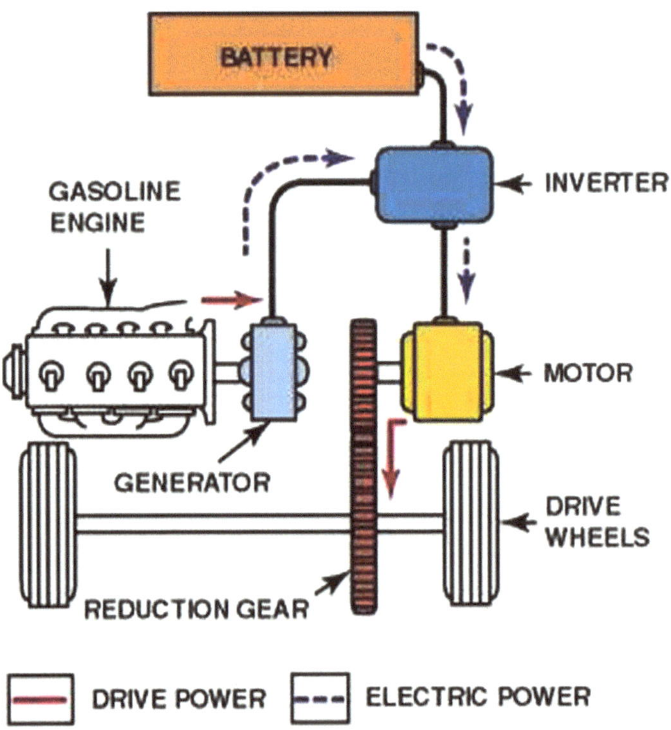

Fig.7.3: Series hybrid

Parallel Hybrid: Both the internal combustion engine and the electric motor are connected to the transmission and can simultaneously drive the vehicle.

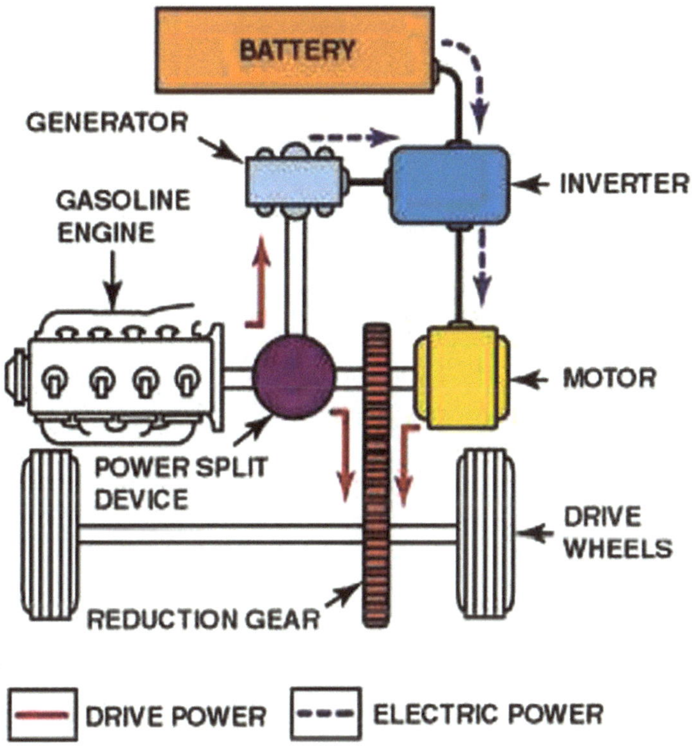

Fig.7.4: Parallel hybrid

Series-Parallel Hybrid: Also known as power-split hybrids, this configuration combines the features of both series and parallel systems, providing more flexibility and efficiency. Many modern HEVs are based on this design.

7.5 Components of a Hybrid Electric Vehicle

Engine: Similar to traditional engines but typically smaller and more efficient. Both petrol and diesel engines are used in HEVs.

Fig.7.5: Engine

Battery: Stores energy generated by the engine or through regenerative braking. Common battery types include Lead Acid (2.2V), Nickel Cadmium (1.2V), and Lithium-ion (3.7V).

Electric Motor: Powers the vehicle at low speeds and assists the engine when additional power is required. Brushless DC motors (BLDC) are commonly used.

Controller: Manages power flow, charging the battery or powering the motor. It converts DC from the battery into modulated DC or AC signals and operates using low-voltage input signals.

Generator: Converts mechanical energy from the engine into electrical energy, which can be used to power the motor or recharge the battery. It also helps in starting the engine quickly.

Fig.7.6: Generator

Advantages of HEVs

Hybrid Electric Vehicles offer several benefits, including fuel savings, low emissions, reduced noise pollution, and lower maintenance requirements. They are also considered safe to drive and represent a significant improvement in fuel efficiency compared to conventional vehicles.

Disadvantages of HEVs

Despite their advantages, HEVs come with certain drawbacks. They are more expensive than conventional vehicles and may carry higher risks in accidents due to their dual fuel system. Parts can be expensive to replace—especially the battery—and these vehicles are often slower and less visually appealing than traditional cars.

Future Scope of HEVs

The future of hybrid vehicles is promising. Major automobile manufacturers are heavily investing in hybrid technology, including for larger vehicles like trucks. Future advancements are expected to enhance vehicle efficiency, incorporate better safety features, increase power output, and reduce overall costs through mass production. As adoption increases, hybrid vehicles will become more affordable and widely available.

Example: Tata's Hybrid Product

Tata Motors has introduced hybrid solutions like the Tata Starbus Hybrid, representing India's commitment to embracing hybrid technology in public transportation.

Tata Motors' Starbus Hybrid represents a landmark in India's journey toward cleaner, more efficient public transportation. Unveiled in the mid-2010s, the Starbus Hybrid is a 12-metre urban bus that combines a BS IV-compliant diesel engine with an electric drive system in a series-hybrid configuration—a first for a locally manufactured hybrid bus in India . Built on Tata's modular Ultra platform and conforming to Urban Bus Specifications (UBS-II), AIS-051 and CMVR norms, it leverages a compact diesel generator to charge on-board lithium-ion phosphate batteries, which in turn power twin electric motors. This layout decouples the engine from the wheels, allowing the diesel unit to operate constantly at its most efficient RPM, while the electric motors deliver smooth, silent propulsion across urban routes.

Powertrain and Performance:

At its heart lies a four-cylinder, turbocharged diesel engine—rated at approximately 195 hp—that drives an alternator rather than the axle directly. Electrical energy is stored in a lithium-ion phosphate battery pack, sized to deliver an effective urban range of up to 250 km under typical operating conditions. The electric drive motors produce instant torque from start-up, eliminating gear-shifting delays, reducing driveline complexity and delivering superior acceleration in stop-and-go city traffic. Regenerative braking captures kinetic energy during deceleration, further enhancing fuel economy and reducing brake wear.

Efficiency and Emissions:

Field trials and early deployments demonstrated that the Starbus Hybrid achieves up to 40 percent better fuel efficiency compared to conventional diesel buses of similar size, translating into substantial operational cost savings for fleet operators. Equally important, tailpipe emissions of particulate matter, nitrogen oxides and carbon dioxide are dramatically lower, making the vehicle particularly attractive to environmentally conscious municipal transport agencies.

Operational Adoption:

The Mumbai Metropolitan Region Development Authority (MMRDA) was the launch customer, placing an initial order for 25 units, with deliveries commencing in April 2017. Priced at ₹2.2 crore ex-showroom, the Starbus Hybrid qualifies for a subsidy of up to ₹61 lakh under India's FAME (Faster Adoption and Manufacturing of Hybrid & Electric Vehicles) scheme, bringing its net cost to around ₹1.6 crore—comparable to premium diesel counterparts. Encouraged by these economics, other city transport undertakings soon followed suit, integrating Starbus Hybrids into their fleets to meet tightening emission norms and passenger comfort expectations.

Passenger Experience and Safety:

Inside, the Starbus Hybrid offers a modern, spacious cabin with large panoramic windows, ergonomic seating, and low-floor accessibility. Wide central aisles and strategically placed grab rails facilitate quick boarding and alighting, reducing dwell times at stops. Safety is augmented by CCTV surveillance, an integrated telematics system for real-time vehicle monitoring, and an Intelligent Transport System (ITS) that provides passengers with accurate estimated-time-of-arrival displays at designated stops.

Strategic Impact and Future Prospects:

As India's cities expand and air-quality concerns intensify, Tata's Star bus Hybrid has set a precedent for sustainable mass transit solutions. Its success has spurred further investment in both hybrid and fully electric buses, encouraging infrastructure upgrades such as depot-based charging systems and renewable-energy integration. Looking ahead, Tata is refining the Starbus Hybrid platform with next-generation battery chemistries and fuel-cell prototypes, aiming to further reduce lifecycle emissions and total cost of ownership—an evolution that underscores India's commitment to green mobility and smart-city initiatives.

India-GTU Questions

Define the term hybridness? chapt7pg220 SUM2025

Chapter 08

Power Electronics in EHV

8 POWER ELECTRONICS IN EHV

8.1 Sizing the power electronics

Sizing power electronics in a hybrid electric vehicle (HEV) involves careful selection of components such as inverters and converters that can effectively handle the vehicle's power demands for driving, regenerative braking, and overall energy management. The process begins with defining system requirements, which includes identifying the power sources and loads by determining the contributions from the internal combustion engine, electric motor(s), and the battery, along with the power needed for auxiliary systems. Establishing operating cycles is equally important, where driving cycles such as the Urban Dynamometer Driving Schedule (UDDS) are analysed to capture the vehicle's speed and load profiles over time. Additionally, performance objectives are set, focusing on goals like maximizing fuel economy, minimizing emissions, or achieving specific acceleration targets.

Once the requirements are defined, the next step is to select key components. Power converters play a central role in this process: DC-DC converters are employed to step up low-voltage DC to the higher-voltage DC bus and to provide power for the 12V auxiliary system, while inverters are responsible for converting DC power from the battery into AC power for the electric motor(s) and for controlling the motor's speed and torque. Alongside converters, the design must also consider energy storage systems by evaluating the power and energy requirements of the battery as well as alternative storage options such as supercapacitors.

Finally, optimization algorithms are utilized to determine the most efficient sizing and configuration of these components. Methods such as dynamic programming and parallel chaos optimization are commonly employed to strike a balance between fuel economy, power density, thermal management, and energy storage capacity. This systematic approach ensures that the power electronics are not only adequately sized but also optimized to meet the overall performance and efficiency goals of the hybrid electric vehicle.

Dynamic Programming (DP) is one of the most widely used methods for sizing power electronics in hybrid electric vehicles, as it determines the optimal energy management strategy (EMS) and component sizes that maximize fuel economy for a given driving cycle. In contrast, rule-based sizing approaches rely on pre-defined rules to size components according to the type of vehicle and its performance requirements. Beyond these, other optimization methods such as parallel chaos optimization are also employed to minimize cost while still meeting the necessary performance constraints.

In addition to optimization strategies, designers must also consider critical constraints and advanced technologies. Thermal management plays a vital role, as effective heat dissipation ensures reliable operation of power electronic components, particularly when advanced semiconductor materials are used. Component costs are another major factor, requiring a balance between performance and affordability, since modern materials like Silicon Carbide (SiC) and Gallium Nitride (GaN) significantly improve efficiency and power density but come with higher initial expenses. Power density and efficiency themselves are key design parameters, as selecting components with higher values allows for smaller, lighter, and more effective systems. Finally, degradation of components such as batteries and fuel cells must be taken into account, with sizing and operation strategies optimized to extend lifespan and maintain long-term performance.

8.2 Requirement of power electronics in HEV

The electrification of the transportation sector is progressing rapidly. Automotive manufacturers are actively implementing comprehensive electrification strategies to meet societal expectations and customer demands by delivering carbon-neutral solutions for the movement of people and goods. Within this transition, power electronics plays a critical role and continues to evolve quickly from a technological standpoint. While several of these technologies have reached a high level of maturity, the resulting systems are often electrically complex and must therefore achieve exceptional reliability. This article concentrates on the reliability of power electronic systems used in electric vehicles (EVs) and hybrid electric vehicles (HEVs). It begins by outlining the reliability requirements and key challenges associated with power electronics in EV/HEV applications. Subsequently, recent advancements in power electronic components are discussed, highlighting how each development contributes to enhancing overall system reliability. The paper also examines reliability-focused design methodologies, illustrated through two practical examples: an onboard charger for EVs and a drivetrain inverter. Finally, the article presents an outlook on future research opportunities related to power electronics reliability in EV and HEV applications. In conclusion, although many reliability-related aspects have already been effectively addressed, the introduction of entirely new technologies presents ongoing challenges, making continued attention to reliability a critical priority.

8.3 Overview of the electrical power system in HEV

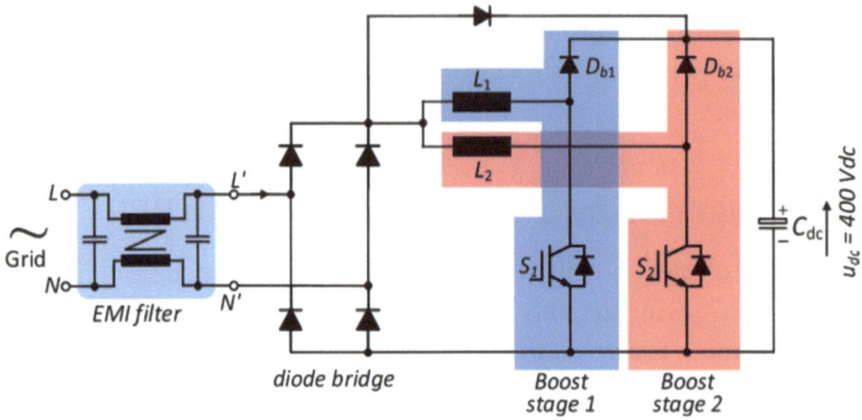

Fig.8.1: Single-phase OBC topology

Fig. shows the most popular single-phase OBC topology, also known as interleaved boost PFC. Although interleaved boost PFC has advantages from efficiency and power density over single-switch boost PFC, it requires higher component count, which may lead to lower reliability. However, since each boosting stage in the interleaved PFC is designed to handle half of the rated power, it can provide redundancy to the system if one boosting stage fails. Therefore, the OBC can continue charging the battery up to half of its rated power. Currently, research and development on more advanced topologies are undergoing, such as totem-pole PFC with wide bandgap (WBG) devices, to further improve the efficiency and the power density of OBC. The long-term trend is moving toward bidirectional charging, where EV/HEV can support the power grid by feeding the battery power back to the grid. One of the One of the primary challenges associated with onboard chargers (OBCs) is reducing charging time. As battery capacities continue to increase to support longer driving ranges, there is a corresponding need for chargers with higher power ratings. Although fast and ultrafast dc charging stations (exceeding 350 kW) can significantly shorten charging durations to approximately 10–15 minutes, their limited availability has led vehicle manufacturers to pursue high-power OBC solutions based on three-phase power supplies (see Fig.).

In most commercially available vehicles, constraints related to power density and physical size restrict three-phase OBC power levels to the range of 11–20 kW. An exception is the Renault Zoe, which employs an integrated motor drive and OBC architecture to achieve a power rating of 43 kW [30]. It is also important to note that the high-voltage dc bus level varies depending on the selected OBC topology.

8.4 Isolated dc-dc converter with unidirectional PSFB configuration

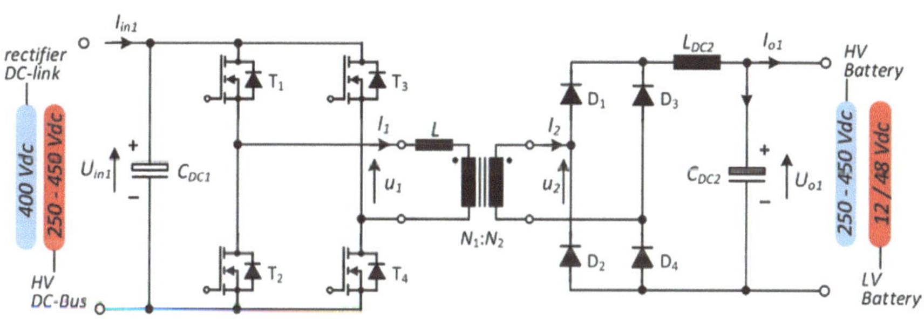

Fig.8.2: Isolated dc–dc converter topologies

The figure illustrates typical isolated dc–dc converter topologies used in electric-based vehicles. These converters are implemented at two distinct stages. The first stage is responsible for charging the battery from the active front-end rectifier, where the DC–DC converter supplies the high-voltage (HV) dc bus, typically operating within a range of 250–450 Vdc. A second dc–dc converter, often based on the same topology, is used to generate the low-voltage (LV) dc bus, which operates at either 12 V or 48 Vdc depending on the vehicle's low-voltage load requirements (see Fig.).

To comply with safety regulations, electrical isolation must be ensured between the power grid and the HV dc bus, as well as between the HV and LV dc buses. As depicted in Fig., this isolation is achieved by dividing the converter into primary and secondary sides through the use of an isolation transformer. The transformer can be implemented using a resonant network, such as a series resonant converter (SRC), commonly referred to as an LLC configuration. Alternatively, as shown in Fig. 3(b), isolation can be achieved by utilizing the transformer's leakage inductance—sometimes supplemented with additional inductance—in a phase-shift full-bridge (PSFB) converter topology.

On the primary side, an H-bridge configuration is typically employed, while the secondary side most commonly uses a diode bridge. Although the diode bridge allows only unidirectional power flow, it simplifies the converter design by eliminating coma plex gate driver requirements, reducing overall cost, and enhancing reliability. However, this approach can negatively impact efficiency due to increased diode conduction losses and limited flexibility in modulation strategies.

The PSFB converter offers favourable operational characteristics, including fixed switching frequency, soft-switching capability, reduced filtering requirements, and suitability for wide power and voltage operating ranges. Resonant-based converters can also deliver high performance when optimized for specific application conditions. Nevertheless, sustaining high efficiency across wide variations in power levels and voltage gain ratios is challenging, particularly due to their variable switching frequency operation. Additionally, designing and tuning the resonant network for high-frequency operation is complex, as it must account for layout constraints, switching devices, transformer characteristics, and parasitic elements. Therefore, the PSFB converter shows promising performance over resonant-based in adapting to different application needs while maintaining a desirable converter performance. This has become a new trend to adapt PSFB converters based on the WBG devices for EV/HEV applications.

8.5 Advances In Power Electronic Components

Advances in wide-bandgap semiconductor materials and power module packaging technologies are key enablers for achieving the development roadmap illustrated in Fig.. Studies have shown that silicon carbide (SiC) MOSFETs using conventional packaging techniques exhibit lower power cycling capability compared with silicon (Si) IGBTs. To address this limitation, numerous advanced packaging concepts have been proposed, as highlighted in the literature, to improve reliability and thermal performance.

The figure presents the research and development roadmap for power electronics in electric-based vehicles, as derived from published reports.

Power electronics R&D requirement	2017		2025		2035		20XX	
	Inverter	OBC/DC-DC	Inverter	OBC/DC-DC	Inverter	OBC/DC-DC	Inverter	OBC/DC-DC
Cost ($/kW)	5	30	4	15	3	8	*> 98% efficient palm size converter*	
Power density (kW/L)	12	2	15	4	60	15		
Efficiency (%)	96	94	97	96	98	97		
	Semiconductor		*Semiconductor*		*Semiconductor*		*Semiconductor*	
Junction Temperature (T_j)	*-40°C to +150°C*		*-55°C to +175°C*		*-55°C to >+175°C*		*-55°C to >+250°C*	

Fig.8.3: EV roadmap year-wise

The figure illustrates the lifetime requirements for automotive applications alongside the research and development trends in power electronics for electric-based vehicles. As indicated, the ongoing demand for higher efficiency and increased power density places increasingly stringent operating requirements on power electronic components used in EV and HEV systems. Meeting these demands necessitates operation at higher power levels, elevated voltages, and increased switching frequencies. When combined with the rigorous operating conditions inherent to automotive environments, these requirements result in particularly harsh conditions for power electronic systems.

Table summarizes the various sources of stress affecting power electronic components and provides representative environmental and stress testing conditions specific to EV and HEV applications. Among the listed factors, thermal stress is identified as one of the most critical contributors to component degradation. This stress manifests both as elevated steady-state operating temperatures and as temperature cycling, both of which significantly impact the reliability and lifetime of power electronic devices.

8.6 Power Electronics in HEV

In hybrid electric vehicles (HEVs), rectifiers convert alternating current (AC) from the generator or the alternating voltage from the battery pack into direct current (DC) to charge the battery. Common rectifier types used in HEVs include diode bridges and controlled rectifiers.

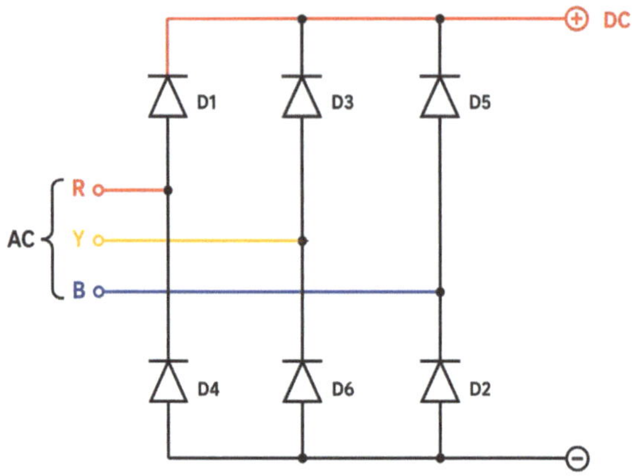

Fig.8.4: Power Rectifier

8.6.1 Rectifiers Used in HEVs

Rectifiers are essential components in Hybrid Electric Vehicles (HEVs) because they convert the alternating current (AC) produced by the electric motor and regenerative braking system into direct current (DC), which is required to charge the battery and operate electronic systems. When the vehicle slows down, the motor functions as a generator, producing AC electricity. Instead of allowing this energy to be lost as heat, rectifiers transform it into usable DC power that can be stored in the battery for future use. They also regulate the distribution of electrical energy among key systems, including the engine, motor, battery, and vehicle accessories such as lighting, infotainment, and air conditioning.

To improve efficiency, modern HEVs employ advanced semiconductor rectifiers made from technologies like silicon carbide (SiC) and insulated-gate bipolar transistors (IGBTs). These materials enable faster switching, lower power losses, and improved thermal performance. As a result, vehicles benefit from better energy management, longer battery life, and increased driving range. Efficient rectification also enhances regenerative braking performance and supports smoother power delivery, which helps reduce fuel consumption and emissions. Overall, rectifiers play a vital role in ensuring efficient power conversion and sustainable operation in hybrid electric vehicles.

8.6.2 Voltage Ripples Buck Converter Used in HEVs

In Hybrid Electric Vehicles, buck converters are widely used to step down high-voltage battery power to lower voltage levels required by auxiliary systems such as lighting, infotainment, and control units. However, one of the key challenges associated with buck converters is voltage ripple, which refers to the small, periodic fluctuations in the output voltage. These ripples are mainly caused by the switching action of power semiconductor devices and the charging and discharging of inductors and capacitors within the converter. Excessive voltage ripple can negatively affect sensitive electronic components, leading to reduced efficiency, electromagnetic interference (EMI), and potential reliability issues. To minimize these effects, HEV power systems use high-quality filters, optimized control strategies, and advanced components such as low-loss capacitors and fast-switching transistors. Reducing voltage ripple improves power stability, protects electronic systems, and enhances overall vehicle performance. Therefore, careful design and control of buck converters are essential to ensure smooth, reliable, and efficient power delivery in hybrid electric vehicles.

8.6.3 Non-Isolated Bidirectional DC-DC Converter

A non-isolated bidirectional DC-DC converter is a power electronic device that enables controlled power flow in both directions between two DC voltage levels without galvanic isolation. In hybrid and electric vehicle applications, it is commonly used to interface the high-voltage battery with the low-voltage DC bus or auxiliary energy storage devices such as supercapacitors. During motoring mode, the converter steps up the low-voltage DC to a higher level to supply the traction motor through an inverter, while in regenerative braking mode, it operates in the reverse direction to step down the voltage and transfer recovered energy back into the battery. Its bidirectional operation is achieved using active switches (typically MOSFETs or IGBTs) arranged in a half-bridge or full-bridge configuration, controlled with pulse-width modulation (PWM) techniques. The absence of a transformer makes the circuit more compact, lightweight, and efficient, although it sacrifices electrical isolation. Key design considerations include voltage conversion ratio, current handling capability, efficiency, and thermal management. In electric and hybrid vehicles, non-isolated bidirectional DC-DC converters are essential for energy management, improving fuel economy, and supporting high power density while maintaining overall system reliability.

8.7 Voltage Source Inverter, Current Source Inverter Isolated Bidirectional DC-DC Converter

A Voltage Source Inverter (VSI) is a power electronic device that converts a fixed DC input voltage into a controlled AC output voltage. The DC side is supported by a large capacitor, which maintains a nearly ripple-free voltage. In operation, semiconductor switches such as IGBTs or MOSFETs are controlled using Pulse Width Modulation (PWM) techniques to generate a sinusoidal AC waveform at variable frequency and amplitude. The VSI is widely used in electric vehicles, particularly for driving Permanent Magnet Synchronous Motors (PMSMs) and Brushless DC motors, since it provides efficient and precise control of motor torque and speed. In addition, VSIs are also applied in on-board charging systems and renewable energy integration.

A Current Source Inverter (CSI), on the other hand, converts a constant DC input current into a controlled AC output current. The DC side of the CSI employs a large inductor to ensure a constant current supply. In operation, the current is switched through semiconductor devices, and commutation circuits are employed to achieve safe switching. Unlike the VSI, where voltage is the controlled parameter, the CSI regulates current and is therefore suitable for applications where load current must be maintained within specific limits. CSIs offer natural short-circuit protection, but they are bulkier due to the large input inductor. In electric vehicles, they are occasionally used for induction motor drives and regenerative braking, although they are less popular than VSIs.

An Isolated Bidirectional DC–DC Converter (IBDC) is an essential interface for energy storage systems in electric and hybrid electric vehicles. It enables bidirectional power transfer between two DC buses of different voltage levels while ensuring galvanic isolation through a high-frequency transformer. During motoring operation, the converter transfers power from the battery to the DC link of the motor drive, whereas during regenerative braking it channels recovered energy from the motor back into the battery. The inclusion of a high-frequency transformer allows both voltage step-up or step-down and electrical isolation, which enhances safety and system flexibility. The IBDC is a key enabler in battery–fuel cell hybrid systems, auxiliary storage integration, and on-board chargers.

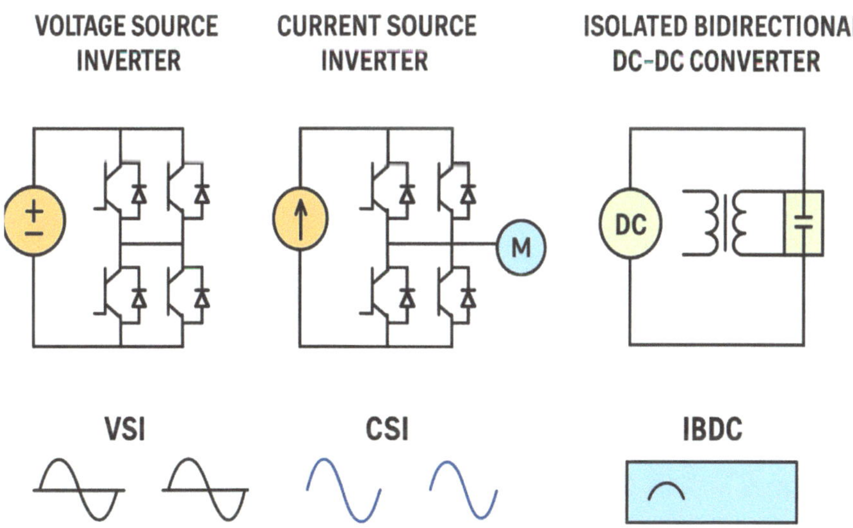

Fig.8.5: Voltage Source Inverter, Current Source Inverter, Isolated Bidirectional DC-DC Converter

8.8 PWM Rectifier in HEV & EV

A PWM Rectifier is a power electronic circuit used in electric and hybrid electric vehicles to convert an incoming AC supply into a regulated DC voltage, while maintaining a nearly sinusoidal input current with unity power factor. Unlike conventional uncontrolled or diode rectifiers, a PWM rectifier employs IGBTs or MOSFETs that are switched using high-frequency Pulse Width Modulation techniques. This active control allows the rectifier to reduce input current harmonics, regulate DC-link voltage, and enable bidirectional power flow between the grid and the vehicle. In electric vehicles, PWM rectifiers are particularly important in on-board charging systems, where they ensure efficient and safe conversion of AC mains power to DC for battery charging. In hybrid electric vehicles, they can also be used to interface auxiliary power units with the DC bus. Additionally, since bidirectional PWM rectifiers can send power back to the grid, they play a key role in vehicle-to-grid (V2G) systems. Thus, PWM rectifiers improve efficiency, reduce Total Harmonic Distortion (THD), and enhance overall power quality in EV and HEV charging and energy management.

8.9 PHEV Battery Chargers

A Plug-in Hybrid Electric Vehicle (PHEV) battery charger is a crucial component that enables energy transfer from the power grid to the vehicle's battery pack. Since PHEVs combine an internal combustion engine (ICE) with an electric drive system, the charger ensures safe and efficient charging of the battery to maximize the electric driving range.

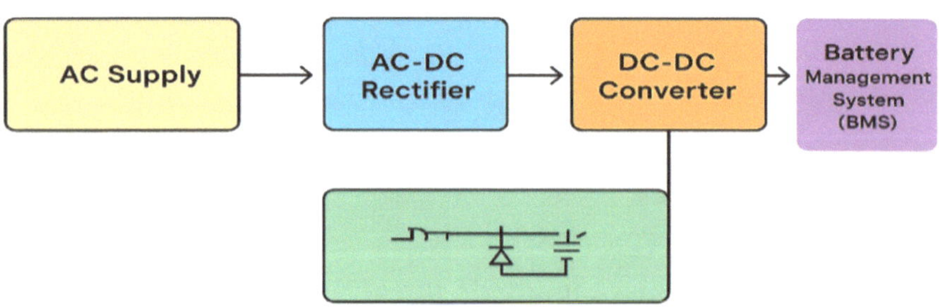

Fig.8.6: PHEV Battery Chargers

8.9.1 Working Principle of PHEV Battery Charger

When a Plug-in Hybrid Electric Vehicle (PHEV) is connected to the power grid for charging, the alternating current (AC) supply from the mains (typically 120 V in the US or 230 V in many other countries) must be converted into a regulated direct current (DC) voltage suitable for charging the traction battery. This conversion is achieved through multiple power electronic stages.

First, an AC-DC rectifier is employed to convert the AC input into an unregulated DC output. Next, a Power Factor Correction (PFC) circuit is incorporated to maintain a near-unity power factor and to minimize harmonic distortions, ensuring efficient energy transfer from the grid. The rectified DC is then processed by a DC-DC converter, which regulates both the voltage and current according to the battery's required charging profile, typically following the Constant Current–Constant Voltage (CC–CV) method.

Finally, the charging process is supervised by the Battery Management System (BMS), which monitors key parameters such as battery voltage, charging current, temperature, State of Charge (SOC), and State of Health (SOH). The BMS prevents overcharging and overheating, and ensures safe, reliable, and efficient battery operation throughout the charging cycle.

There are two main types of chargers:

- **On-board Chargers (OBC):** Compact units integrated inside the vehicle (3.3–22 kW), mainly used for slow (Level 1) and moderate (Level 2) charging.
- **Off-board Chargers:** High-power units installed at charging stations (up to 350 kW) that enable DC fast charging, bypassing the OBC.

Charging is classified into three levels: Level 1 (slow AC charging), Level 2 (fast AC charging), and Level 3 (DC fast charging), with charging times ranging from several hours to less than 30 minutes.

Modern chargers also support bidirectional power flow for Vehicle-to-Grid (V2G) applications, improving grid stability and energy management.

8.9.2 PHEV parallel configuration

In a parallel configuration of a Plug-in Hybrid Electric Vehicle (PHEV), both the internal combustion engine (ICE) and the electric motor are mechanically coupled to the transmission system and can directly deliver power to the wheels. This arrangement allows the vehicle to operate in multiple modes, such as electric-only drive, engine-only drive, or a combined hybrid drive.

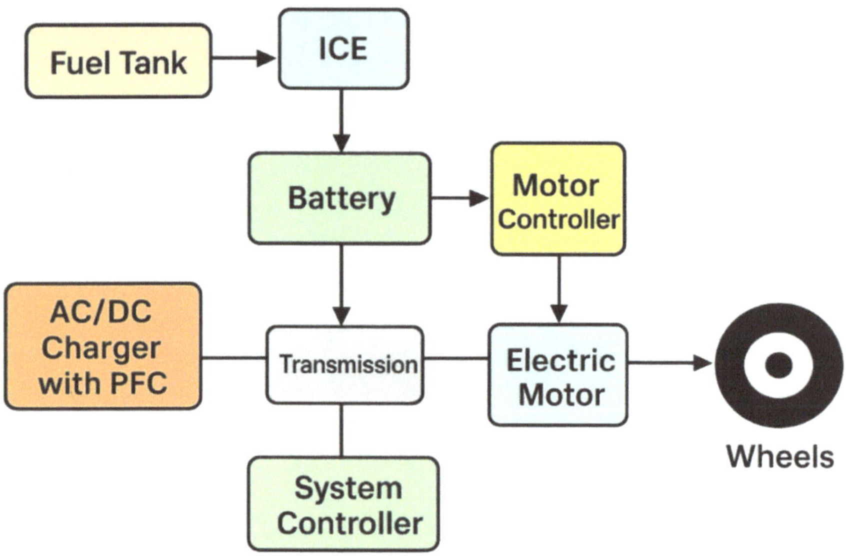

Fig.8.7: PHEV parallel configuration

The high-voltage battery supplies power to the electric motor through a motor controller and inverter, while the AC/DC charger with power factor correction (PFC) handles grid charging. The ICE uses fuel from the fuel tank and is connected to the wheels via a clutch and transmission system. A system controller coordinates energy flow, ensuring optimal performance, fuel economy, and reduced emissions. Regenerative braking enables the motor to act as a generator, feeding energy back into the battery. The parallel configuration provides high flexibility and efficiency since either source can propel the vehicle independently or jointly, depending on driving conditions. This makes PHEVs suitable for urban driving on electric mode and highway driving with the ICE, while also allowing grid charging to extend the driving range.

Chapter 09

Advancement in EV

9 ADVANCEMENT IN EV

9.1 ADAS

Advanced Driver Assistance Systems (ADAS) in electric vehicles (EVs) are electronic technologies that use a network of sensors—such as LiDAR, radar, cameras, and ultrasound—to enhance vehicle safety and efficiency. In 2025, ADAS will have become a standard benchmark for modern EVs, providing both safety-critical interventions and comfort-focused automated features. The core Advanced Driver Assistance Systems (ADAS) features expected in 2025 electric vehicles include Adaptive Cruise Control (ACC), which maintains a pre-set distance from the car ahead by automatically controlling speed. Automatic Emergency Braking (AEB) is another key feature, detecting imminent collisions with objects or pedestrians and applying brakes if the driver fails to react. Lane Keeping Assist (LKA) & Lane centering uses cameras to detect road markings and provides steering input to keep the vehicle within its lane. Additionally, Blind Spot Detection (BSD) alerts the driver to vehicles in areas not visible in mirrors, and Traffic Sign Recognition (TSR) identifies road signs like speed limits and displays them on the dashboard or adjusts vehicle speed automatically.

Smart Grid technology

In 2026, the evolution of electric mobility has reached a milestone where Advanced Driver Assistance Systems (ADAS) and Smart Grid infrastructure operate as a singular, interconnected network. Rather than focusing solely on safety, current ADAS platforms leverage Vehicle-to-Grid (V2G) protocols to transform EVs into mobile energy storage units that automatically stabilize local power networks during peak demand. These vehicles utilize "energy-intelligent" navigation, which cross-references real-time grid load data with sensor-based traffic analysis to suggest routes that maximize regenerative braking and prioritize charging stations powered by renewable sources. By integrating V2X (Vehicle-to-Everything) communication, the car's automated systems can modulate charging speeds or discharge power back to the home, ensuring that the vehicle's energy consumption is perfectly synchronized with the stability and availability of the broader electrical ecosystem.

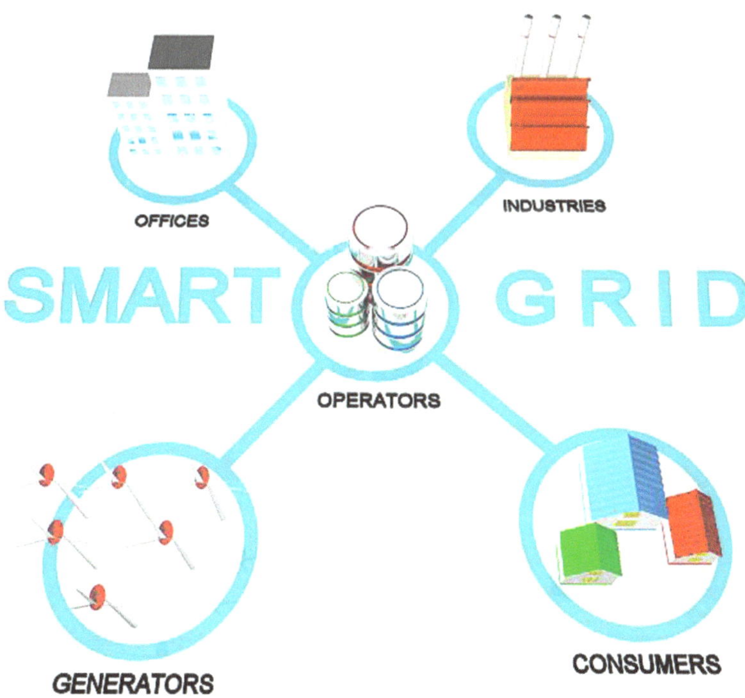

Fig.9.1: Smart grid operation

9.2 Role of AI in EV

Artificial Intelligence is revolutionizing electric vehicles by making them smarter, safer, and more efficient. AI powers autonomous driving, enabling cars to interpret surroundings and make real-time decisions. It enhances battery management systems, predicting battery health and optimizing charging cycles for longer life. Through predictive maintenance, AI detects early signs of faults, reducing breakdowns and costs. It also improves energy efficiency by managing power distribution between the motor, battery, and regenerative braking. Finally, AI connects EVs to smart grids, allowing intelligent charging and even vehicle-to-grid energy sharing, paving the way for sustainable mobility.

1. Autonomous EV on a smart road
2. AI-powered battery management visualization
3. Smart grid with EVs charging intelligently

Intelligent EV charging with a smart grid uses two-way communication to manage charging times, shifting demand from peak hours to off-peak, balancing grid load, and integrating renewables for efficiency and cost savings, preventing overloads while ensuring EVs are charged when needed, often using Demand-Side Management (DSM) and Vehicle-to-Grid (V2G) tech.

Key Benefits include Reduced Grid Strain, which prevents blackouts by flattening peak demand and avoiding expensive upgrades; Cost Savings, through lower electricity bills for users via time-of-use pricing; Renewable Integration, which maximizes the use of solar/wind power when available, reducing fossil fuel reliance; Grid Resilience, where V2G allows EVs to send power back to the grid during outages, acting as mobile batteries; and Future-Proofing, as new features and demands can be updated remotely. Technologies involved consist of Smart Chargers, connected devices that modulate power flow; Demand-Side Management (DSM), programs to shift loads; Vehicle-to-Grid (V2G), where EVs supply power back to the grid; and Artificial Intelligence (AI), used for prediction and optimization.

9.3 Battery technology

9.3.1 Solid-state battery

Solid-state batteries (SSBs) are poised to revolutionize EVs by replacing flammable liquid electrolytes with solid materials, promising huge leaps in safety, energy density (longer range), faster charging, and longer lifespan, essentially eliminating range anxiety and fire risk, though challenges in mass production and cost remain, with major players like Toyota, BMW aiming for limited commercial release around 2027-2028 to usher in a truly transformative era for electric mobility.

Fig.9.2: Solid state battery

9.3.2 Advantages and Industry Impact

Solid-state batteries (SSBs) are poised to transform the EV market by replacing the flammable liquid electrolytes in current lithium-ion (Li-ion) batteries with a non-flammable solid material, such as a ceramic, glass, or polymer. This fundamental change offers significant performance benefits:

- Higher Energy Density: SSBs can achieve an energy density of over 400 Wh/kg (with prototypes reaching 600 Wh/kg), potentially doubling the range of current EVs to over 1,000 km on a single charge.
- Enhanced Safety: The use of solid electrolytes dramatically reduces the risk of thermal runaway and fires, a major safety concern with Li-ion batteries.
- Faster Charging: The design allows for faster movement of ions, with the potential for ultra-fast charging (e.g., 10-80% in 10-15 minutes), comparable to refueling a gasoline car.
- Longer Lifespan: SSBs are more resistant to degradation over time, promising a longer overall battery lifespan, potentially lasting over a decade.

Major Players and Timelines

Numerous companies are in a global race to commercialize solid-state battery technology, with significant progress made in 2025 in moving from lab prototypes to vehicle testing.

- Toyota: A frontrunner in R&D, Toyota plans to launch its first EV with solid-state batteries as early as 2027-2028, targeting a range of 1,000 km.
- Quantum Scape: Backed by Volkswagen, this US-based startup is developing solid-state lithium-metal batteries and is shipping advanced prototype cells to partners for testing, with mass production expected by 2030.
- BMW: In collaboration with Solid Power, BMW is road-testing an i7 prototype featuring all-solid-state battery cells to evaluate durability and performance.

- Nissan: Nissan plans to launch a pilot factory in 2024 and aims for mass production of in-house developed solid-state batteries by 2028.
- CATL & BYD: These Chinese battery giants are investing heavily in both semi-solid and all-solid-state solutions, with plans for small-batch production and vehicle integration around 2027.

Remaining Challenges

Despite rapid progress, several hurdles must be overcome before widespread mass-market adoption:

- Manufacturing Costs: Current production methods are complex and expensive, making SSBs two to three times costlier than Li-ion batteries.
- Scalability: Scaling up production from pilot lines to meet the massive demand of the global EV market remains a significant challenge.
- Technical Hurdles: Issues like ensuring a stable interface between solid components, preventing lithium dendrite formation, and optimizing performance across a wide range of operating temperatures are ongoing areas of research

9.3.3 Battery swapping

Battery swapping in Electric Vehicles (EVs) is a system where a depleted battery is quickly exchanged for a fully charged one at automated stations, offering near-instant "refueling" (minutes vs. hours), reducing downtime for fleets (taxis, delivery), lowering EV upfront costs via Battery-as-a-Service (BaaS), and easing grid load by charging batteries off-peak, though it faces challenges like battery standardization and infrastructure cost.

Fig.9.3: Battery swapping

9.4 Wireless charging lane -Pilot project

Norway has implemented a groundbreaking **wireless charging road system** that enables electric vehicles (EVs) to recharge their batteries while in motion. This technology, also referred to as an «electric road system» (ERS), aims to eliminate the need for traditional charging stops and improve the practicality of electric mobility.

The system works by using special copper coils beneath the road surface that generate an electromagnetic field, which is then picked up by a receiver on the underside of a compatible EV, converting the energy into electricity to charge the battery. This dynamic charging method reduces charging downtime, which is particularly beneficial for commercial fleets like taxis and buses that operate continuously throughout the day. By enabling on-the-go charging, the technology also reduces the need for large, heavy batteries, thereby decreasing vehicle weight and increasing energy efficiency.

Norway has launched pilot projects in cities like Trondheim and Oslo to test this technology in real-world conditions, including harsh Scandinavian winters, with plans to expand the infrastructure to over 1,000 kilometres of highway by 2030. These initiatives involve collaborations with international partners, including the Israeli tech firm Electron Wireless, semiconductor manufacturers like Infineon, and vehicle manufacturers such as Yutong and Higer. The projects support

Norway's ambitious goal for 100% of all new car sales to be zero-emission by the end of 2025.

Fig.9.4: Electric Road Systems (ERS): Paving the Way for a Sustainable Future in Transportation

The global push for green and efficient transportation systems has given rise to Electric Road Systems (ERS), also referred to as *Charge As You Drive (CAYD)* or *Electric Highways*. These innovative systems allow vehicles to charge dynamically while in motion, offering a groundbreaking solution to the challenges posed by electrifying heavy-duty trucks and commercial fleets. With pilots underway in countries like France, Sweden, and Belgium, ERS is poised to revolutionize transportation infrastructure, drastically reducing carbon emissions and dependency on static charging stations.

This comprehensive analysis delves into ERS technology, its global applications, benefits, challenges, and its potential to transform the future of logistics and freight mobility.

ERS is a transportation system where electric vehicles (EVs) charge directly from the road through embedded infrastructure. Unlike traditional EVs that rely solely on static charging points, ERS-equipped vehicles can recharge their batteries while driving. This technology primarily caters to commercial vehicles such as trucks, buses, and freight carriers, addressing their need for longer operational ranges and reducing downtime for recharging.

Three key ERS technologies are currently being piloted globally:

1. Inductive Charging: Wireless charging via coils embedded beneath the road.
2. Conductive Charging: Contact-based charging through rails installed on the road surface.
3. Overhead Catenary Systems: Power lines with pantographs for trucks and buses.

Global ERS Initiatives and Pilots

1. France: Leading the ERS Revolution

   France has taken a pioneering role in developing ERS technology through multiple pilot projects:

   - A10 Highway Pilot: In collaboration with Vinci, Elanroad, and the Université Gustave Eiffel, this pilot integrates inductive charging systems into the A10 highway. Objective: Enhance energy efficiency for heavy vehicles and reduce dependence on massive static charging networks.
   - eRoadMontBlanc: Another initiative led by the Université Gustave Eiffel, this project focuses on testing electrified road solutions in the Chamonix-Mont Blanc valley, a region with high traffic and environmental sensitivity.

2. Belgium: Pioneering Urban Freight Electrification

   In Brussels, a partnership between AISIN and Elanroad is evaluating conductive charging technologies for urban and regional freight networks.

- Key Goal: Assess large-scale deployment feasibility for dynamic charging systems in densely populated urban areas.

3. Sweden: A Global Leader in ERS

   Sweden has emerged as a trailblazer in ERS deployment, having conducted successful pilots that have set benchmarks for the global community:

   - eRoadArlanda: A 2 km pilot project where conductive rails embedded in the road charged EVs dynamically.
   - Gotland Pilot: Focused on testing wireless inductive charging for buses and heavy trucks. Results: Reduced battery sizes by 20%-30%, significantly cutting vehicle costs. Achieved 80% reduction in CO_2 emissions compared to diesel-powered vehicles.

Sweden aims to expand its ERS network to over 3,000 km of highways by 2040, showcasing its commitment to achieving carbon neutrality in freight transport.

The benefits and challenges of Electric Road Systems (ERS) highlight its potential to revolutionize long-haul logistics. Below is the conclusion of the points you began: Challenges and Limitations (Continued) 3. Maintenance and Durability (Continued) Road-embedded systems face constant exposure to extreme weather, heavy axle loads, and debris. Maintaining electrical integrity in diverse climates (e.g., snow, heavy rain, or extreme heat) requires specialized materials and frequent inspections, which can increase long-term operational expenditures. 4. Regulatory and Legal Frameworks Implementing ERS across borders requires complex legal agreements regarding energy taxation, billing for international fleets, and road safety regulations. Determining who owns the infrastructure—governments or private utilities—remains a significant hurdle for policymakers. 5. Vehicle Retrofitting Requirements While ERS reduces battery size, it requires vehicles to be equipped with specialized pick-up equipment (pantographs, conductive arms, or inductive plates). The initial cost of retrofitting existing electric or diesel fleets may deter smaller operators until the technology reaches mass-market scale. The Path Forward For ERS to succeed in 2025 and beyond, several strategic steps are essential: Public-Private Partnerships (PPP): Collaborative funding models can distribute the high "per-kilometer" cost between governments and logistics giants. Pilot Corridors: Expanding successful trials, such as those in Sweden (eRoadArlanda) and Germany, into permanent "electric backbones" for major freight routes. Unified Standards: Developing international technical standards to ensure a truck manufactured in one country can seamlessly draw power from the roads of another. By addressing these infrastructure and standardization gaps, ERS can transform from a series of pilot projects into a cornerstone of global decarbonized transport.

9.5 Onboard charger

An Onboard Charger (OBC) in an electric vehicle is the built-in power electronics module that converts AC electricity from the grid into DC electricity for the battery. It determines the maximum AC charging rate (typically 3.6–22 kW) and ensures safe, efficient charging under the supervision of the Battery Management System (BMS). An Onboard Charger (OBC) is a device permanently installed inside electric vehicles that serves as the critical interface between external AC power sources, such as wall outlets or public charging stations, and the vehicle's DC battery system. Its core function is to convert Alternating Current (AC) from the grid into Direct Current (DC), which the battery can store for later use. Beyond simple conversion, the OBC plays an essential role in charging by negotiating parameters with the charging station, rectifying current, and ensuring the battery's health and longevity throughout the process. Key features of the OBC include its ability to safely transform household or public AC power into the precise voltage and current required by the battery, its definition of the maximum AC charging rate (typically ranging from 3.6 to 22 kW depending on design), and its close integration with the Battery Management System (BMS) to monitor safety, charging curves, and efficiency. By enabling convenient charging at home, workplaces, and destinations without relying solely on DC fast chargers, the OBC makes everyday EV use practical and accessible.

Comparison: Onboard Charger vs. DC Fast Charger

Here's the comparison table between **Onboard Chargers (OBC)** and **DC Fast Chargers** for electric vehicles:

Table 9.1: Comparison of chargers

Feature	**Onboard Charger (OBC)**	**DC Fast Charger**
Power Source	AC from grid/outlets	DC directly supplied
Conversion Location	Inside the EV	Outside (charger unit)
Charging Speed	3.6–22 kW (slower)	50–350 kW (much faster)
Use Case	Home/workplace charging	Highway, long trips
Cost & Complexity	Lower cost, lighter	Higher cost, bulky

CHARGER RENTAL SERVICE

EV charger rental services provide temporary access to charging equipment, whether stationary or mobile, without requiring users to purchase or install their own chargers. These services often include mobile DC fast chargers that can be delivered directly to a customer's location, short-term rentals tailored for events, construction sites, or fleet operations, and turnkey solutions that cover installation, maintenance, and technical support. This approach is part of a broader trend known as Charging as a Service (CaaS), in which third-party providers take responsibility for owning, operating, and maintaining the charging infrastructure, offering users a flexible and cost-effective way to meet their EV charging needs.

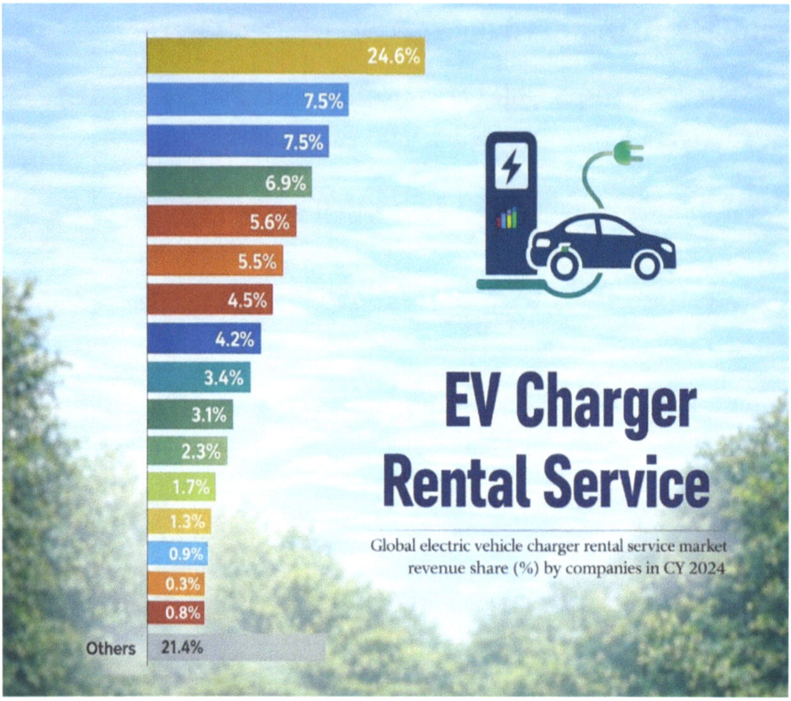

Fig.9.5: Charger rental service survey for different companies

9.6 China Revolution

China has emerged as the global leader in the revolution of electric vehicles (EVs) and electric hybrid vehicles (EHVs) through a combination of strong government policies, subsidies, battery innovation, and supply chain dominance. Early on, the Chinese government identified EVs as a strategic industry and supported their growth with massive incentives, tax breaks, and strict regulations on fuel consumption and emissions, which accelerated adoption. At the same time, China gained control over the battery supply chain, from raw materials like lithium and cobalt to cell manufacturing, with companies such as CATL and BYD becoming world leaders in battery production. By 2024, China produced 12.4 million EVs—about 70% of global output—and nearly half of all new cars sold domestically were electric, showing how mainstream adoption has become. Five of the world's top ten EV manufacturers are Chinese, including BYD, SAIC, and Geely, and the country is rapidly expanding exports to Europe, Southeast Asia, and Latin America with affordable models that are reshaping global markets. This aggressive expansion has triggered trade tensions with the US and EU, but it underscores the scale and speed of China's transformation from a follower in the automotive industry to the epicentre of electrified mobility. While challenges such as domestic market saturation, overcapacity risks, and international trade barriers remain, China's EV and EHV revolution has fundamentally reshaped the global auto industry and forced other nations to rethink their industrial strategies and competitiveness.

The electric vehicle (EV) revolution has achieved global leadership by integrating aggressive industrial policy, a dominant battery supply chain, and massive infrastructure scaling. In 2024, China accounted for over 50% of global EV sales and produced nearly 30% of the world's vehicles.

Market Leadership & Dominance (2024–2025)

- Sales Milestones: EV sales in China surpassed 50% of the passenger vehicle market for the first time in 2024. By 2025, new targets aim for a significant portion of all newly added vehicles to be New Energy Vehicles (NEVs).
- Domestic Giants: BYD has overtaken Tesla as the world's largest EV manufacturer by sales volume. Other major domestic players include NIO, XPeng, and Geely.
- Export Surge: China is the world's largest vehicle exporter, projecting over 7 million units by the end of 2025, with EVs driving roughly 40–50% of this tota.

Key Drivers of the Revolution

1. Strategic Government Support: Since 2009, the "Made in China 2025" plan has provided an estimated $60 billion in subsidies, tax breaks, and R&D funding. While direct consumer purchase subsidies ended in 2022, non-monetary incentives like green license plate privileges and parking fee exemptions continue to drive demand.
2. Battery Ecosystem Control: China controls over 70% of global lithium battery production and leads in mining and processing critical minerals like lithium, cobalt, and nickel. CATL alone maintains a ~37% share of the global EV battery market as of 2025.
3. Infrastructure Scale: China hosts the world's largest charging network, with over 2 million public and private charging points. New targets aim for 28 million charging facilities by 2027 to support up to 80 million EVs.
4. Technological Innovation: Beyond traditional BEVs, China is a leader in battery-swapping technology (led by NIO and Aulton) and extended-range electric vehicles (EREVs) that offer combined ranges exceeding 2,100 km.

Emerging Trends for 2025

- Shift to Hybrids: To counter range anxiety and international tariffs, manufacturers are shifting focus toward Plug-in Hybrids (PHEVs), which saw a 210% surge in exports in early 2024 compared to 40% for pure battery vehicles.
- Global Manufacturing: Facing rising EU and US tariffs, Chinese automakers are building local factories in Hungary, Turkey, Thailand, and Brazil to maintain market access.
- "China Speed" in Development: Chinese firms can develop and launch new models in just 18 months, compared to the five-year cycles typical in Japan or the US

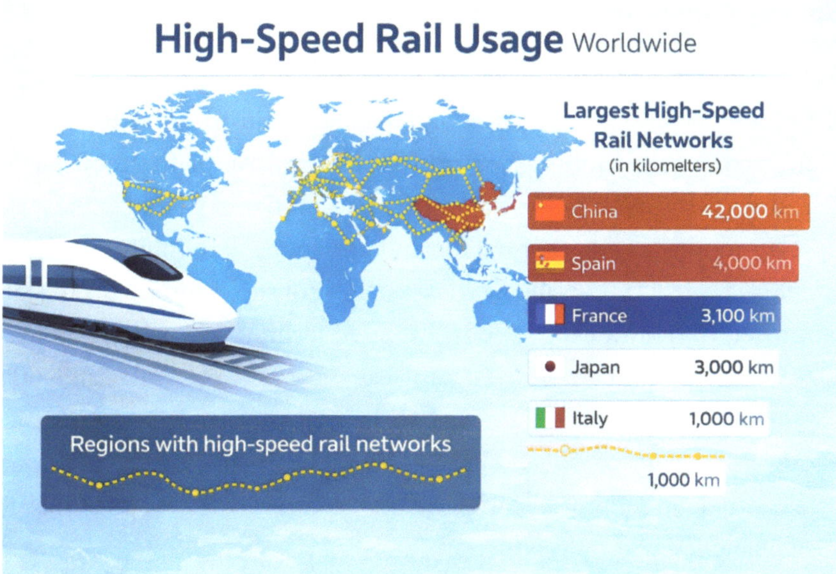

Fig.9.6: High-speed rail network

9.7 Energy consumption by dew point dehumidifier

In EV battery manufacturing, specifically within "dry rooms" where moisture-sensitive lithium is handled, dew point dehumidifiers are critical for quality and safety and represent a major portion of the total energy consumption.

Role in Manufacturing

Moisture Prevention: Lithium is highly hygroscopic; any contact with moisture can lead to faulty batteries, reduced performance, and potential fire or explosion risks. IN ultra-Low Dew Points, dehumidifiers maintain extremely low humidity levels in dry rooms, typically below 1% Relative Humidity (RH). Maintaining these conditions ensures product integrity, optimal electrode performance, and consistent manufacturing processes.

Energy Consumption and Efficiency

Major Energy User: The dry room conditioning (including dehumidification) is the single largest energy consumer in battery cell manufacturing, accounting for up to 43% to 60% of the total energy demand of the plant. Energy Efficiency Solutions: Manufacturers are developing advanced, energy-efficient dew point dehumidifiers. Some systems can reduce overall electrical energy consumption by up to 85% compared to traditional methods. Patented technologies, like the "Green Dry Purge," recycle waste heat and can save between 30% and 60% in energy use. Dual-rotor designs with internal heat recovery can cut energy use by up to 50%. The high energy demand of the dry room makes efficient dehumidification solutions crucial for cost-effective and sustainable EV battery production.

Chapter 10

Alternative Sources Hydrogen Fuel

10 ALTERNATIVE SOURCES OF HYDROGEN FUEL

Hydrogen cannot be considered a primary fuel because it is not naturally available in a ready-to-use form. It is created from hydrogen gas through processes that require energy input from other sources, such as electricity or fossil fuels. This means para-hydrogen does not produce energy by itself. Instead, it acts as an energy carrier, storing the energy used during its production and releasing it when needed for scientific, industrial, or storage purposes. Its role is to transport and deliver energy efficiently, rather than serving as an original energy source.

10.1 History of Hydrogen

- **1766 – Henry Cavendish**

  In the year 1766, British chemist Henry Cavendish discovered a new type of gas that burned readily, which he referred to as "inflammable air." Through careful experiments, he observed that when this gas reacted with oxygen during combustion, water was formed as a product. This finding was important because it challenged the earlier belief that water was a basic element. Instead, Cavendish's work suggested that water was created from the reaction of this gas, later known as hydrogen, with oxygen. His discovery played a key role in advancing the scientific understanding of chemical elements.

- **1783 – Antoine Lavoisier**

  In 1783, the French chemist Antoine Lavoisier made a major contribution to the study of hydrogen. He introduced the name "hydrogen," derived from the Greek words *"hydro," meaning water, and "genes,"* meaning creator, because the gas produces water when it burns. Lavoisier also demonstrated through experiments that water is not an element but a compound formed from hydrogen and oxygen. This discovery helped establish the modern understanding of chemical composition and played an important role in the development of modern chemistry.

- **20th Century**

  During the 20th century, hydrogen became an important fuel for space exploration, especially in the form of liquid hydrogen used in rocket engines. Its high energy content made it ideal for launching spacecraft. As technology advanced, hydrogen also began to attract attention as a clean and efficient energy carrier. Researchers explored its potential for use in vehicles and electricity generation, since it produces only water as a by-product. This shift marked the beginning of hydrogen's role in supporting environmentally friendly energy systems for transportation and power production.

10.2 Green Hydrogen: Clean Energy Loop

Sunlight or wind generates electricity → electricity splits water → hydrogen stores energy → hydrogen releases energy with water as the only byproduct, which is both elegant and climate-aligned.

Renewable-powered water electrolysis is emerging as a highly sustainable and technically sound method for producing hydrogen, as it relies exclusively on electricity generated from solar, wind, or hydropower sources to split wate (H_2O) into hydrogen (H_2) and oxygen (O_2), resulting in a hydrogen supply with no direct greenhouse gas emissions. The fundamental mechanism of this process is electrochemical in nature: when direct current is applied across two electrodes immersed in water with an electrolyte, hydrogen is formed at the cathode through reduction reactions, while oxygen is released at the anode via oxidation reactions, following the overall reaction $2H_2O \rightarrow 2H_2 + O_2$.

Fig.10.1: Hydrogen Train

This process is implemented using different types of electrolyzers, including alkaline, proton exchange membrane (PEM), and solid oxide electrolyzers, each offering distinct advantages in terms of efficiency, operating temperature, and compatibility with variable renewable power. Among these, PEM electrolyzers are particularly effective when integrated with solar and wind systems because they can respond rapidly to fluctuating electricity supply, which is crucial for maintaining system stability.

From an energy standpoint, the production of one kilogram of hydrogen through electrolysis typically requires approximately 50–55 KWh of electrical energy, while the energy content of the hydrogen produced is about 33.3 kWh, indicating that efficiency improvements remain an active area of research and development. When renewable electricity is used, the resulting hydrogen is classified as green hydrogen, setting it apart from conventional hydrogen produced using fossil fuels. This green hydrogen has wide-ranging applications across multiple sectors, including transportation through fuel cell vehicles, industrial processes such as ammonia synthesis and steel manufacturing, and large-scale energy storage systems where surplus renewable electricity can be converted into hydrogen and stored for later use.

A significant advantage of renewable-based electrolysis is its contribution to energy system flexibility, as it enables the conversion of intermittent renewable energy into a stable and transportable fuel. Conceptually, the process forms a closed and environmentally benign cycle in which renewable electricity produces hydrogen, hydrogen is used to deliver energy, and water is regenerated as the only byproduct. Several regions with abundant renewable resources are already deploying pilot and commercial-scale projects, including electrolyzers directly connected to offshore wind farms and utility-scale solar installations. Although challenges such as high capital investment, material degradation, and water availability must still be addressed, ongoing technological advancements and cost reductions are steadily improving feasibility. Overall, water electrolysis powered by renewable energy stands out as a key enabler of deep decarbonization and the long-term transition to a low-carbon energy economy.

10.3 Physical and Chemical Properties in Renewable-Based Water Electrolysis

10.3.1 Physical Properties Relevant to Electrolysis

Water (H_2O)

- Physical state: Liquid at room temperature (25°C)
- Density: ~1000 kg m^{-3} at 4°C
- Electrical conductivity: Pure water is a poor conductor; electrolytes (KOH, NaOH, or proton-conducting membranes) are required to enhance ion transport.
- Boiling point: 100°C (important for system temperature limits in alkaline and PEM electrolyzers)

Hydrogen (H_2)

- Physical state: Colourless, odourless, tasteless gas
- Density: 0.0899 kg m^{-3} at STP (lightest known gas)
- Energy content:
  - Gravimetric: ~120 MJ kg^{-1} (very high)
  - Volumetric: ~10.8 MJ m^{-3} (low, hence compression or liquefaction needed)
- Diffusivity: Very high, enabling rapid mixing but requiring leak-resistant storage systems
- Storage conditions:
  - Compressed gas: 350–700 bar
  - Liquid hydrogen: –253°C

Oxygen (O_2)

- Physical state: Colourless gas
- Density: 1.429 kg m^{-3} at STP
- Often released to the atmosphere or captured for medical and industrial use

10.3.2 Chemical Properties Governing Electrolysis

Electrochemical Nature

Water electrolysis is a non-spontaneous redox reaction driven by electrical energy.

Cathode Reaction (Reduction):

$$2H_2O + 2e^- \rightarrow H_2 + 2OH^-$$

Anode Reaction (Oxidation):

$$4OH^- + O_2 \rightarrow 2H_2O + 4e^-$$

Overall Reaction:

$$2H_2O\ (l) + 2H_2\ (g) + O_2\ (g)$$

Minimum theoretical voltage required:

- 1.23 V at 25 °C
- Practical systems operate at 1.8–2.2 V due to overpotentials and losses.

3. Chemical Properties of Hydrogen

   - Highly reactive with oxygen:

Fuel Cell Reaction Example:

10.3.3 Effect of Renewable Electricity on Properties

Renewable electricity does not alter the chemical properties of hydrogen; however, it ensures zero CO_2 emissions during production, significantly lowers environmental impact, and enables the classification of the produced hydrogen as sustainable or "green hydrogen."

Key difference between Physical & Chemical properties of Hydrogen

Table 10.1: Properties of Hydrogen

Substance	Key Physical Property	Key Chemical Property
Water	Poor conductor (pure)	Undergoes electrolysis
Hydrogen	High energy per kg	Clean combustion
Oxygen	Supports combustion	Oxidizing agent

10.4 Hydrogen Storage for Automobile Applications (I.C. Engine & Fuel Cell Vehicles)

Effective hydrogen storage is a key enabler for the adoption of hydrogen-fueled I.C. engine and fuel cell vehicles. While hydrogen offers an outstanding gravimetric energy density of nearly 120 MJ/kg, ranking it among the most energy-dense fuels, its very low volumetric energy density at normal conditions creates significant storage challenges. To carry sufficient fuel for practical driving ranges, hydrogen must be stored in ways that do not compromise vehicle size, safety, or efficiency. Consequently, automotive hydrogen storage systems must be compact, lightweight, safe, and economically viable, while complying with strict performance and safety standards. To address these requirements, advanced storage approaches such as compressed hydrogen tanks, cryogenic liquid hydrogen storage, and metal hydride-based systems are employed, making hydrogen-powered automobiles technically feasible.

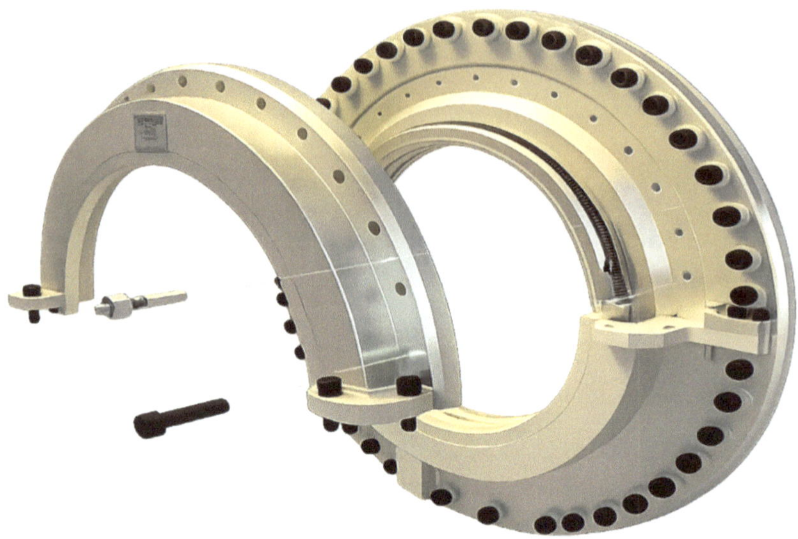

Fig.10.2: Hydrogen seal ring

Advantages, Limitations, and Automobile Use:

Compressed hydrogen storage offers several advantages for automotive applications due to its simple and well-established technology, enabling reliable operation and easy integration into vehicle designs. One of its major benefits is rapid refuelling, typically completed within 3–5 minutes, making it comparable to conventional fuel refilling. In addition, the presence of a mature and expanding infrastructure supports its practical deployment in hydrogen-powered vehicles. However, this storage method also has notable limitations, including low volumetric energy density, which necessitates larger storage volumes, and heavy high-pressure tanks that add to vehicle weight. Furthermore, the hydrogen compression process requires significant energy, leading to an energy loss of approximately 10–15%. Despite these challenges, compressed hydrogen storage is successfully used in commercial fuel cell vehicles such as the Toyota Mirai and Hyundai NEXO, demonstrating its current viability for automobile applications.

Estimation of Hydrogen Stored in a High-Pressure Tank

Numerical 1. Estimate the mass of Hydrogen considering a hydrogen storage tank with a volume of 120 liters operating at a pressure of 700 bar, which is commonly used in modern fuel cell vehicles. At this pressure, the density of compressed hydrogen is approximately 42 kg/m^3.

Solution: convert the tank volume from liters to cubic meters:

$$120\text{" L"} = 0.12\text{" " "}m^3\text{"}$$

The mass of hydrogen stored in the tank can be calculated using the relation:

$$\text{"Mass of hydrogen"} = \text{"Density"} \times \text{"Volume"}$$

$$\text{"Mass"} = 42\text{" "\{"kg/m"\}}\text{^}3 \times 0.12\text{" ""m" ^}3 = 5.04\text{" kg"}$$

Thus, a 120-liter tank at 700 bar can store approximately 5.04 kg of hydrogen. This quantity of hydrogen is sufficient to provide a driving range of about 500–600 km for a typical hydrogen fuel cell vehicle, highlighting the practical feasibility of high-pressure compressed hydrogen storage for automotive applications.

Numerical 2: The metal hydride storage system has a hydrogen storage capacity of 1.5 wt%. Calculate the mass of hydrogen that can be stored in 100 kg of the metal hydride material. Comment on the suitability of this storage system for modern automobile applications.

Solution:

The mass fraction of hydrogen stored in the metal hydride is given as 1.5 wt%, which is equivalent to 0.015.

$$\text{"Mass of hydrogen"} = 0.015 \times 100 = 1.5\text{" kg"}$$

The metal hydride system can store approximately 1.5 kg of hydrogen per 100 kg of material.

10.5 Storage Selection for Automobile Applications

The selection of an appropriate hydrogen storage system is closely linked to the specific requirements of the vehicle technology. In hydrogen-fueled I.C. engine vehicles, compressed hydrogen storage is commonly adopted because it is a well-established method that can be easily incorporated into conventional engine designs. These engines are comparatively tolerant to fuel impurities and can withstand severe combustion environments, which makes high-pressure gaseous storage a reliable and practical choice. For fuel cell vehicles, compressed hydrogen storage at 700 bar has become the preferred option in commercial applications, as it provides an effective compromise between driving range, safety standards, and rapid refuelling capability. While liquid hydrogen storage offers the prospect of increased energy density, its use remains largely confined to research and prototype vehicles due to high energy consumption during liquefaction and the need for complex cryogenic systems. Likewise, metal hydride-based storage, although inherently safe and capable of high volumetric hydrogen storage, is currently unsuitable for automotive use because of its excessive weight and slow hydrogen absorption and desorption rates.

Comparison of Hydrogen Storage Methods:

Table 10.2: Storage Methods for Hydrogen

Parameter	Compressed	Liquid	Metal Hydride
Gravimetric Density	Medium	High	Low
Volumetric Density	Low	Medium	High
Safety	High	Medium	Very High
Cost	Medium	High	High
Vehicle Use	✔✔✔	✔ (limited)	✗ (R&D)

10.6 Hydrogen Vehicles

Hydrogen vehicles are already in use in the form of fuel cell electric vehicles (FCEVs) and experimental hydrogen I.C. engine vehicles. A well-known commercial example is the Toyota Mirai, which uses compressed hydrogen stored at 700 bar to generate electricity through a fuel cell, producing only water as exhaust. Another prominent example is the Hyundai NEXO, a hydrogen fuel cell SUV that offers a long driving range and fast refuelling time. Additionally, the Honda Clarity Fuel Cell is an example of a hydrogen-powered passenger car developed to demonstrate the feasibility of fuel cell technology. These vehicles illustrate the practical application of hydrogen as a clean automotive fuel.

Adoption of hydrogen fuel cell buses in public transport

India has recently taken significant steps toward the adoption of hydrogen fuel cell buses in public transport, reflecting its commitment to clean and sustainable mobility. In June 2025, NTPC Limited handed over five hydrogen fuel cell buses to the Leh Administration, marking India's first commercial deployment of hydrogen buses. These buses operate from a dedicated green hydrogen mobility station and are specifically engineered to perform under harsh, sub-zero climatic conditions at an altitude of approximately 11,562 feet, demonstrating the feasibility of hydrogen mobility in remote and high-altitude regions while substantially reducing carbon emissions.

In another major initiative, the Yamuna Expressway Industrial Development Authority (YEIDA) has proposed the operation of hydrogen buses along the Noida–Agra corridor (about 200 km), utilizing hydrogen supplied from NTPC's green hydrogen facility. These buses are expected to offer long driving ranges, along with modern features such as air-conditioning and GPS tracking, to support sustainable intercity transportation. Furthermore, under the National Green Hydrogen Mission, several pilot projects involving hydrogen buses and other hydrogen-powered vehicles have been launched across different regions of the country to develop clean mobility infrastructure and evaluate large-scale fuel cell vehicle deployment. In addition, hydrogen bus demonstration runs in Delhi, including those attended by international delegations, have showcased India's technological progress and growing global interest in hydrogen-based public transport. Collectively, these initiatives highlight India's transition from pilot testing to early commercial adoption of hydrogen fuel cell buses as part of its broader strategy to decarbonize the public transport sector.

List of Figures

List of Tables

Paper Solution

India-GTU QUESTION PAPER - SUM2025

BE -SEMESTER –VII EXAMINATION –SUM2025

Subject Code: 3170923 | **Date: 08-05-2025**

Subject Name: | **Electrical and Hybrid Vehicle**

Time: 02:30 PM TO 05:00 PM | **Total Marks: 70**

Instructions:

1. **Attempt all questions.**
2. **Make suitable assumptions wherever necessary.**
3. **Figures to the right indicate full marks.**
4. **Simple and non-programmable scientific calculators are allowed.**

			Marks
Q.1	**(a)**	Give Comparisons of Hybrid Electric Vehicles and Conventional Vehicles? chapt2pg36	**03**
	(b)	Explain different types of forces acting on vehicle due to air friction. chapt1pg11,12,14	**04**

	(c)	Prepare a block diagram of Battery Electric Vehicle (BEV) with all necessary components and explain it in brief. chapt1pg8	**07**
Q.2	**(a)**	List economic and environmental impact of electric hybrid vehicle. chapt2 pg52	**03**
	(b)	Give the operation and controlling of DC motor drive. chapt3pg72	**04**
	(c)	Compare series hybrid and parallel hybrid system with their merits and demerits. chapt3pg49	**07**
		OR	
	(c)	Explain historical development of automobile and development of interest and activity in the EV from 1890 to present day. chapt1pg3	**07**
Q.3	**(a)**	What type of operation is known as flux weakening in permanent magnet machines? chapt3pg94	**03**
	(b)	Draw torque-speed characteristics of following motors and suggest best for EV with reason. (1) chapt3pg85 Induction Motor (IM) (2) Permanent Magnet Synchronous Motor (PMSM) (3) Switched Reluctance Motor (SRM) (4) chapt3pg91 Permanent Magnet Brushless DC Motor (PMBLDC)chapt3pg84pg60	**04**

	(c)	Explain the difference between ultracapacitor and battery as an energy Storage device for EV.chapt4pg118	**07**
		OR	
Q.3	**(a)**	Explain following terms (1) Vehicle to Grid (V to G) (2) Vehicle to Load (V to L) (3) Vehicle to Vehicle (V to V) chapt7 pg218	**03**
	(b)	List the optimization-based strategies in Hybrid Electrical Vehicles.chapt6pg182	**04**
	(c)	ExplainElectronically controlled regenerative brakingsystem functioning as an ABS.chapt5pg54	**07**
Q.4	**(a)**	What are the main issues with fuel cells in electric vehicle application? chapt1pg74	**03**
	(b)	Prepare schematic diagram of solar and grid connected charging station. chapt6pg186	**04**
	(c)	Prepare and explain a complete driving control method for Induction Motor in electrical vehicle application. chapt3pg88	**07**
		OR	
Q.4	**(a)**	Define the term hybridness? chapt7pg220	**03**
	(b)	What is a plug-in HEV? Explain with figure. chapt2pg40	**04**
	(c)	Define following terms (1) Specific energy (2) Specific power (3) Energy density (4) Ambient temperature (5) Life cycle of battery (6) Energy stored (7) Charge capacity chapt4pg113	**07**

Q.5	**(a)**	Give the Differences Between BEV, HEV and FCEV. chapt5pg147	**03**
	(b)	Draw and explain driver circuit for 3 phase SRM-switched reluctance motor drive. chapt3pg91	**04**
	(c)	Draw and Explain typical CAN network for HEVs. chapt2pg56	**07**

OR

Q.5	**(a)**	Explain the Energy Management concept for HEV. chapt6pg177	**03**
	(b)	Draw and explain driver circuit for 3 phase BLDC motor. chapt3pg66,68	**04**
	(c)	Give the Comparison of Lithium Ion and Lead Acid Battery Configurations in EV. chapt4pg66, 68	**07**

India-GTU QUESTION PAPER - WIN2024

BE -SEMESTER –VII (NEW) EXAMINATION –WIN2024

Subject Code:3170923 **Date:19-11-2024**

Subject Name: **Electrical and Hybrid Vehicle**

Time:10:30 AM TO 01:00 PM **Total Marks:70**

Instructions:

1. **Attempt all questions.**
2. **Make suitable assumptions wherever necessary.**
3. **Figures to the right indicate full marks.**
4. **Simple and non-programmable scientific calculators are allowed.**

Marks

Q.1	**(a)**	What is the need and importance of electric and hybrid electric vehicles? chapt1 pg 31pg24	**03**
	(b)	With the help of a neat block diagram explain different subsystems of electric drive train. chapt3 pg62,79	**04**
	(c)	Derive equations & explain different types of forces acting on vehicle going uphill with neat diagram. chapt1 pg14	**07**
Q.2	**(a)**	Compare hybrid electric vehicles with conventional vehicles. chapt1pg36,37	**03**
	(b)	Write two merits and two demerits of electric vehicles. chapt1pg7	**04**
	(c)	Draw and explain the architecture and power flow control of a parallel hybrid electric drive train. chapt2pg50,51	**07**
		OR	
	(c)	(i) Compute aerodynamic drag, rolling resistance & gradient force for the following vehicle assuming ρ = 1.2 kg/m3 & θ = 8°. Gross vehicle weight = 1500 kg, CD=0.3, Area = 2.5 m2, μ=0.015, v = 80 kmph, Tyre radius = 0.3 m. chapt3pg16 (ii) Also compute total traction force assuming pick-up from 0 to 50 kmph in 20 sec, with linear acceleration & zero slope.	**07**
Q.3	**(a)**	What is a plug-in HEV? Explain with figure. chapt4 pg40	**03**
	(b)	What is meant by (i) Series hybrid electric vehicle. (ii) Parallel hybrid electric vehicle.chapt2 pg38	**04**

	(c)	Define the terms charge capacity, specific energy, energy density, specific power, charge efficiency, energy efficiency, Life cycle of battery. chapt4 pg113	07
		OR	
Q.3	(a)	Give comparisons of Hybrid Electric Vehicles and Conventional Vehicles.chapt2 pg36	03
	(b)	List the various components of HEV drive train. chapt3pg64 Mention the advantages of hybrid vehicles. Chapt3pg66	
	(c)	With neat diagram explain the construction and working of lead-acid battery. chapt4pg121	07
Q.4	(a)	Classify the electric motors drives for EV and HEV application. chapt3 pg70	03
	(b)	List the optimization-based strategies in Hybrid Electrical Vehicles? chapt6pg182	04
	(c)	Explain the basic principle of Super Capacitors based Energy Storage System in Hybrid Electric Vehicles? chapt4pg116	07
		OR	
Q.4	(a)	What is the advantage of AC motor over DC motors for EV applications? chapt1 pg23	03
	(b)	Give a classification of different energy management strategies. chapt6pg203	04
	(c)	Explain the fuel cell and flywheel as energy source elements in electric and Hybrid electric vehicles. chapt5pg131,136	07
Q.5	(a)	List the various components of HEV drive train. chapt3 pg64	03

	(b)	Briefly explain construction and working of SRM. Chapt 3 pg91	**04**
	(c)	Discuss various electric drive train topologies. chapt3 pg44,178	**07**

OR

Q.5		Which types of qualities are required in electric motors for EV application? chapt3 pg55	**03**
		Draw only torque-speed characteristics of following motors. chapt3 pg84 (1) DC Shunt Motor (2) Permanent Magnet Synchronous Motor (PMSM) (3) Permanent Magnet Brushless DC Motor (PMBLDC).	**04**
		Explain the two-quadrant operation of chopper DC motor drive with suitable waveforms for electric vehicle. chapt4pg137	**07**

India-GTU QUESTION PAPER - WIN2025

BE -SEMESTER –VII EXAMINATION –WIN2025

Subject Code:3170923 **Date: 13-11-2025**

Subject Name: **Electrical and Hybrid Vehicle**

Time:10:30 AM TO 01:00 PM **Total Marks: 70**

Instructions:

1. **Attempt all questions.**
2. **Make suitable assumptions wherever necessary.**
3. **Figures to the right indicate full marks.**

4. Simple and non-programmable scientific calculators are allowed.

			Marks
Q.1	**(a)**	List the Economic and environmental impact of electric & hybrid vehicles. pg52	**03**
	(b)	Give a detailed classification of EVs used in the current Automobile market.	**04**
	(c)	Give Comparisons of Hybrid Electric Vehicles and Conventional Vehicles. pg 36,37	**07**
Q.2	**(a)**	Explain the series and parallel configuration of EHV.	**03**
	(b)	Electric Vehicle (EV) is the future of transportation? Justify your answer.	**04**
	(c)	Derive the vehicle dynamic equation and explain all acting forces on the vehicle with appropriate figures. pg11,12,14	**07**
		OR	
	(c)	Explain power flow in Parallel Hybrid HEVs. pg50,51	**07**
Q.3	**(a)**	Explain DC motor speed control techniques.	**03**
	(b)	Explain the classification of Electric Drives used in EVs.	**04**
	(c)	Draw and explain a Hybrid Electric Vehicle (HEV) with an appropriate block diagram.	**07**
		OR	
Q.3	**(a)**	Explain regenerative braking in EV and HEV.	**03**
	(b)	List major advantages of permanent magnet machines in Electrical Vehicles.	**04**
	(c)	Draw and explain different driving cycles. How did it help in the modelling of EVs?.	**07**

Q.4	(a)	Classified different Energy Storage devices used in Hybrid and Electric Vehicles	**03**
	(b)	Explain the History of EVs and HEVs.	**04**
	(c)	Compare Lead-acid battery and nickel-cadmium battery.	**07**

OR

Q.4	(a)	Explain Flywheel in detail. pg 131	**03**
	(b)	Explain different control techniques of Induction Motor drives. pg 88	**04**
	(c)	Explain Solar Powered EHVs.	**07**
Q.5	(a)	Classified different energy management strategies. pg203	**03**
	(b)	Explain the energy-saving potential of hybrid drive trains.	**04**
	(c)	Explain Metal Air Batteries in Details.	**07**

OR

Q.5	(a)	Explain the Fuel Cell EV.	**03**
	(b)	Explain the Basic Principle of Super Capacitors. pg 116	**04**
	(c)	Define following terms	**07**

(1) Specific energy

(2) Specific power

(3) Energy density

(4) Ambient temperature

(5) Life cycle of battery

(6) Energy stored

(7) Charge capacity

Bibliography & References

Proceedings of the IEEE (Volume: 90, Issue: 2, February 2002)

https://www.slideshare.net/slideshow/unit4-power-electronics-in-hevspdf/261703858#4

VYAS INSTITUTE OF ENGINEERING AND TECHNOLOGY, JODHPUR

International Conference on Advanced Mechanical and Industrial Engineering, IOP Conf. Series: Materials Science and Engineering 909 (2020) 012025 IOP Publishing doi:10.1088/1757-899X/909/1/012025

Hybrid Vehicle Energy Model: Analysis of Gas-Electric Hybrid Fuel Efficiency Greg Zilberbrant, McMaster School of Engineering Practice, August 2008

Electric Vehicle Technology Explained-James Larminie, *Oxford Brookes University, Oxford, UK* John Lowry *Acenti Designs Ltd., UK*

Modelling, design and control of a light electric vehicle with hybrid energy storage system for Indian driving cycle-S Devi Vidhya and M Balaji-Measurement and Control 2019, Vol. 52(9-10) 1420–1433_ The Author(s) 2019 Article reuse guidelines: sagepub.com/journals-permissio nsDOI:10.1177/0020294019858212 journals.sagepub.com/home/mac

https://electricvehicles.in/lead-acid-versus-lithium-ion-battery/

AllCell Technologies LLC

https://mechtex.com/blog/working-of-bldc-motor

https://www.circuits-diy.com/how-to-build-a-3-phase-brushless-bldc-motor-driver-circuit/

NPTEL – Electrical Engineering – Introduction to Hybrid and Electric Vehicles; Joint initiative of IITs and IISc

https://www.slideshare.net/slideshow/unit4-power-electronics-in-hevspdf/261703858#4

https://www.google.com/url?sa=i&url=https%3A%2F%2Fcities-today.com%2Fbelgian-cities-invest-electric-hybrid-bus-fleets%2F&psig=AOvVaw3MCQxm_TNhRuLlsibpFkgL&ust=1756877450868000&source=images&cd=vfe&opi=89978449&ved=0CBgQjhxqFwoTCNCji9qsuY8DFQAAAAAdAAAAABAE

Figure referred from the dissertation of the University of Michigan

World Electrical. Veh. J. 2024, *15*(9), 424; https://doi.org/10.3390/wevj15090424

S Devi Vidhya, Department of Electrical and Electronics Engineering, SSN College of Engineering, Kalavakkam 603 110, Tamil Nadu, India.

International Conference on Advanced Mechanical and Industrial Engineering

Measurement and Control 2019, Vol. 52(9-10) 1420–1433

https://www.istockphoto.com/photo/a-hydrogen-fuel-cell-bus-gm1320833040-407254284

https://innovationspace.ansys.com/courses/courses/ev-subsystem-design-of-drive-train/lessons/subsystem-design-of-ev-drive-train-i-lesson-1/

https://onlinecourses.swayam2.ac.in/nou24_ec10/preview

https://archive.nptel.ac.in/content/syllabus_pdf/108103009.pdf

https://nptel.ac.in/courses/108103009

https://archive.nptel.ac.in/content/storage2/courses/108103009/module3/lec8/1.html

https://nptel.ac.in/courses/108106170

https://archive.nptel.ac.in/content/storage2/courses/108103009/module3/lec5/1.html

https://www.lhpes.com/blog/what-are-electric-vehicle-subsystems-and-their-interactions

https://innovationspace.ansys.com/courses/courses/electric-and-hybrid-vehicles/lessons/electric-vehicle-systems-and-configurations-lesson-1/

https://en-roads.climateinteractive.org/scenario.html?v=26.3.0&p1=40&p16=-0.05&p373=50&p375=45&p61=10

https://footprint.conservation.org/en-gb/

https://www.footprintcalculator.org/home/en

https://emissioncalculator.ecotransit.world/

www.ingramcontent.com/pod-product-compliance
Lightning Source LLC
LaVergne TN
LVHW052149210926
848482LV00019B/300